THE MONARCH BUTTERFLY MIGRATION

THE MONARCH BUTTERFLY MIGRATION

Its Rise and Fall

MONIKA MAECKLE

UNIVERSITY OF OKLAHOMA PRESS : NORMAN

Publication of this book is made possible through the generosity of Edith Kinney Gaylord.

Library of Congress Cataloging-in-Publication Data

Names: Maeckle, Monika, 1956– author.
Title: The rise and fall of the monarch butterfly migration / Monika Maeckle.
Description: Norman : University of Oklahoma Press, [2024] | Includes bibliographical references and index. | Summary: "Traces the history of the monarch butterfly migration through the people, issues, and politics surrounding their preservation"— Provided by publisher.
Identifiers: LCCN 2023056020 | ISBN 978-0-8061-9456-1 (hardcover)
Subjects: LCSH: Monarch butterfly—North America. | Monarch butterfly—Migration— North America. | Monarch butterfly—Conservation—North America. | Monarch butterfly—Effect of human beings on—North America.
Classification: LCC QL561.N9 M293 2024 | DDC 595.78/9—dc23/eng/20240208
LC record available at https://lccn.loc.gov/2023056020

The paper in this book meets the guidelines for permanence and durability of the Committee on Production Guidelines for Book Longevity of the Council on Library Resources, Inc. ∞

1   2   3   4   5   6   7   8   9   10

For all my favorite creatures, especially Bob,
Nicolas, Alexander, Opa, and Oma

▼ ▼ ▼

CONTENTS

▼ ▼ ▼

PREFACE

Butterflies have captivated humans for time immemorial, but when *National Geographic* announced the "discovery" of the monarch butterfly overwintering sites in the Mexican mountains on its August 1976 cover, a particular orange-and-black butterfly was thrust into the spotlight. In the ensuing decades, monarch butterflies captured the imagination of the world, especially those of us who live along their migratory flyway, which spans a trinational corridor touching Canada, the United States, and Mexico.

How and why have we become so captivated by this insect? In some ways the answer is obvious: they are beautiful and accessible, they don't sting or bite, and they set examples of extreme resilience and transformation. Yet, despite our deep and genuine appreciation for butterflies and monarchs in particular, we can't seem to agree on how best to support them. Why? And when scientific studies conflict with firsthand experience and conservation advocacy, how do each of us decide the right thing to do?

These questions have driven me for two decades and prompted an exploration for answers through a website I launched called the Texas Butterfly Ranch. It began as a personal blog in 2010 and tool for learning about new media (my profession) and butterflies (a newfound passion). That simple act brought about a midlife metamorphosis that turned me into a full-time pollinator advocate.

The journey has taken me from the northern tip of the Great Lakes to the remote mountains of the Trans-Mexican Volcanic Belt, and many points in between. I've stopped in at Monsanto in St. Louis for a tour, spent time at Monarch Watch in Kansas, visited the roosting sites in Mexico and California multiple times, and interviewed hundreds of professional and citizen scientists, policymakers, and butterfly enthusiasts. I've also spent the better part of three decades in a part of the world where monarchs flit along our San Antonio River much of the year and form impressive overnight roosts in the Texas Hill Country in the fall. And as founder of San Antonio's Monarch Butterfly and Pollinator Festival, I've had the privilege of hosting some of the top monarch scientists in person, in my hometown.

Not long after I started on this curiosity tour, queries asking, "Where is the Texas Butterfly Ranch?" began to appear regularly in my online mailboxes, most often when monarch butterflies were on the move in spring and fall. People were intrigued by my chronicles of caterpillar wrangling and kayak tagging along the Llano River, the discovery of the monarch roosting sites, and the conflicting guidance they found online regarding raising monarchs and particular milkweeds. They wanted to come visit the Texas Butterfly Ranch in person. "Is this a real place? I don't see an address," readers would message me.

Physically, the Texas Butterfly Ranch encompasses the geographic area around Austin, San Antonio, and the Texas Hill Country—the famous "Texas Funnel" through which all monarch butterflies east of the Rocky Mountains pass during their spectacular fall migration. It includes an amazing pollinator garden in downtown San Antonio, a place where I grow mostly native plants, occasionally collect eggs and caterpillars of monarchs and other butterflies, and cultivate host and nectar plants in my *mariposario,* a stacked-rock butterfly house designed by our architect son, Nicolas Rivard. It also encompasses a wildlife-riddled stretch of the Llano River in the Texas Hill Country.

Metaphorically, I like to say the Texas Butterfly Ranch is a lens through which to view complex issues—migration, immigration, climate change, and sustainability.

When my family acquired our ranch on the Llano River in 2000, it took five years for me to realize that native milkweeds populated the riverbanks. Later, when I realized the pecan trees on the bank opposite our house served as roosts to thousands of monarch butterflies each fall, that was it. I was hooked. What's most striking in hindsight: *they had always been there, I just hadn't noticed.* My hope is that this book will nudge readers into noticing, too.

THE MONARCH BUTTERFLY MIGRATION

INTRODUCTION

Every Super Bowl Sunday the United States imports about 65 tons of avocados so we can enjoy our guacamole while rooting for our favorite football team and passing judgment on the year's best television ads. An estimated 80 percent of avocados consumed in the United States come from Mexico—the state of Michoacán, to be exact, which happens to be the only location on earth where avocado trees can bloom 365 days a year, according to the Avocado Institute of Mexico.

Until recently, Michoacán was the only state in Mexico allowed to export avocados into the United States legally. Jalisco was authorized to send the luscious fruits our way in the summer of 2022. The explosive U.S. demand for avocados has created a multibillion-dollar export market, earning the lucrative superfood the nickname *oro verde*, or "green gold" in Mexico.

A more sinister label: "blood fruit." The avocado's profit potential has led Mexican crime cartels increasingly to insert themselves into the avocado business, which has resulted in the clear-cutting of high-altitude forests to make room for more avocado farms. Reports of intimidation and extortion, narco tariffs, farm seizures, "disappearances," and murders of growers are not uncommon. News accounts describe ongoing encroachment of *los carteles* on Mexico's 30,000 avocado growers—the vast majority of whom run small family farms of less than five acres.

So what does any of this have to do with monarch butterflies? Turns out the Goldilocks climate and rich, volcanic soil of Michoacán create not only the perfect ecosystem for year-round avocado production but also the ideal landscape for the unique forest where monarch butterflies have chosen to overwinter for centuries, perhaps millennia. Each fall, mega-millions of monarch butterflies migrate to a particular stretch of avocado-friendly habitat, located at about 10,000 feet altitude, seventy-five miles west of Mexico City. There, they pause all reproductive activities in a semi-hibernative state until spring, when they rouse to continue the life cycle.

Should we stop serving the holy sacrament of guacamole on Super Bowl Sunday to help conserve the precious landscape that hosts the monarch butterfly roosting sites? That's a complicated question and one of many explored in this book. You likely think of monarch butterflies as charismatic creatures that embody tenacity, beauty, and magic. They also provide an accessible window into the avocado quandary. Mexican farmers need to feed their families. Resisting cartels can be deadly. We love avocados, and our support of the avocado industry helps pay the bills in Michoacán—even though avocado farming is undermining the forests where monarch butterflies roost. Is our gusto for guacamole and avocado toast helping or hurting?

Simple answers are evasive and spawn more questions. Take science, for example. As a writer and journalist, I had always considered science to be objective, fact based, and unassailable. But diving into the milkweeds and the subject of monarch butterflies has given me pause. In journalism, we say that "news" is whatever the editor says it is. In science, it often seems no different, only that those crafting the chosen narrative are armed with PhDs.

Anyone can go to Google Scholar and find a research paper that seems to support just about any point of view they choose to espouse. Monarch butterfly experts often disagree. Sometimes they even overlook the facts in the name of a greater good—like when monarch butterfly pioneer Fred Urquhart deliberately misled the public about the roosting site's location "until it was protected." Or when renowned lepidopterist Lincoln Brower lent his name to the petition soliciting "threatened" status for monarch butterflies under the Endangered Species Act. When asked if he really believed that monarch butterflies were threatened with extinction, Brower responded, "Don't you think it's done some good?"

When a lack of consensus occurs in the scientific community, laypeople are often told "the science is in process," but when corporations like Monsanto/Bayer employ scientists to fashion what they brand as "our science" public skepticism results. Nonscientists are left wondering not only who, but what, to believe.

As the founder of San Antonio's Monarch Butterfly and Pollinator Festival, I've navigated myriad questions about the appropriateness of planting non-native milkweeds or using commercially reared butterflies for our annual educational celebration. The facts and opinions on both sides of these arguments are compelling.

And the questions are never-ending. Recent studies suggest that 40 percent of insect species are in decline worldwide.[1] Fewer butterflies and other insects exist than ever, so does that make it okay to breed monarch butterflies in your kitchen or buy commercially raised butterflies from professional breeders to provide a learning opportunity for those with little access to nature? Doing so can inspire a lifelong appreciation for the ecosystem workers that sustain us, but it can also result in spreading disease to the very creatures we aspire to protect. It's clear that our affection for this mesmerizing insect sometimes blinds us to the unintentional harm we can cause by overlooking its wildness and needs beyond the human realm.

And, as much as we remember the good ol' days when butterflies, fireflies, and roly-polies occupied children more readily than virtual experiences via screens, the world is changing. So is the monarch butterfly and its migration.

What to believe? What's the right thing to do? After years of observation, research, and reporting, I'm finding that universally "right" answers don't always exist. As my history-loving son Alexander Rivard likes to say, it's more nuanced. This book explores those nuances and encourages readers to do the same.

This is an unconventional narrative. While it generally moves chronologically through time, the book tells the story in the style of a patchwork quilt. Each chapter addresses a particular aspect of the monarch butterfly's place in our world through a researched, sometimes personal narrative. Read in succession and stitched together with firsthand experience and anecdotes, the goal is to present a larger, more holistic view of our special relationship with *Danaus plexippus*—from the politics, the professional science, and the citizen science to the people devoted to monarchs. This book is shared from the perspective of a decades-long personal fascination with this creature, which began one day on the Llano River in the glorious Texas Hill Country.

▼ ▼ ▼

CHAPTER 1

MAGICAL MONARCHS
ON THE LLANO

It's dry and windy and hotter than the average October day when I launch my kayak on the Llano River about an hour before sunset in search of monarch butterflies. Cicadas chitter as I push off from the put-in spot. A long, spotted gar flushes in the shallow water.

On this Saturday in 2005, a late-season nectar buffet awaits the migrating insects—just the fuel needed to power their flight to Mexico. Our family ranch in the Texas Hill Country sits in the heart of the "Texas Funnel," the channel through which all migrating monarch butterflies east of the Rocky Mountains pass on their route to and from their winter roosts in Mexico.

My friend Jenny Singleton had introduced me to tagging monarch butterflies the previous fall. We tagged them at her place, in nearby Hext, an exercise I found enchanting. Today I was betting the butterflies would be present on our stretch of the Llano, given that we had a similar ecological set-up—a river, flowers, pecan trees nestled against a protective limestone escarpment. I had bought one hundred tags from Monarch Watch and set out this afternoon, searching.

Would I find them?

I paddle past stands of goldenrod toward the Big Riffle, our preferred wading spot, as it thrums a comforting score. A few cardinal flowers flaunt their delicate red petals, and occasional purple asters peek from the tall grasses that grace the Chigger Islands.

Monarch butterflies drift in a steady stream around and across the river, mostly from our side to the opposite. I scoot against the current along the karst-riddled bed toward three low-hanging pecan trees, their bare limbs stooped in submission to repeated floods. Small, bushy cedars line the river bottom and release a piney scent. A belted kingfisher flies ahead, annoyed by my intrusion. He releases his rattling call.

I step onto the bank and approach the pecan trees. Gnats hum in my ears as my red rubber boots stick a moment in the mud. I look up.

A silent eruption of monarch butterflies wafts from the earth. Hundreds of them drift skyward, each of my steps disrupting their peace. The orange-and-black creatures lilt and glide with the rugged Llano Uplift as their backdrop. Floating, flitting, fleeting—they alight on flowers and find their way to the trees.

I grab my net and started swooping. Usually I nab just one or two, but one swing nets twenty-two butterflies. I kneel down, lay my gear on the ground, and swaddle the net's white mesh upon itself to calm them. When I reach in to retrieve one, I remember plunging my hand into a minnow bucket on my father's fishing boat as a child. The captives flutter and wriggle, the teeming epitome of life.

One by one, I place a small, round tag on each butterfly's discal cell, the pale gold center of gravity on the monarch's wing. The strategic placement of the tag allows the butterfly to continue its flight unencumbered—hopefully, all the way to Mexico.

I record the date, time, tag number, and insect's sex. Then comes the best part—opening my soft grasp as the butterfly takes flight. No matter how many times I do it, a smile sprawls across my face.

Buen viaje, mariposa monarca! Have a great trip!

Within fifteen minutes, I run out of tags. I kayak back to the house, returning shortly with another sheet. Upstream, another pecan stand shades a thick frostweed patch. Monarchs perch like tiny birds on the white flowers, sipping nectar. By dusk I tag fifty monarchs.

The following Monday I return to work, but I can't stop thinking about monarch butterflies. I'm compelled to research their biology and natural history. They haunt me—in a good way. I felt their velvety softness. I marveled at their beauty and appreciated their resilience. I pressed the tags securely to their wings with a gentle but firm grasp. And then I released them to the wind for their final eight-hundred-mile flight "home" to a place they've never been.

CHAPTER 2

MONARCH BUTTERFLY, GATEWAY BUG

Eat nonstop for weeks. Poop as needed. Outgrow your flashy caterpillar suit five times. Do the evocative twisty dance. Transform into a jade green chrysalis with a fancy gold necklace, then fall into a deep sleep. During your chrysalis coma, convert legs into wings. One random morning, burst from your shell, drop your damp wings, wait for them to dry, then fly off to have sex with multiple partners in faraway places before returning to earth to die. That's pretty much the compelling story of the monarch butterfly life cycle. From a public relations perspective, it doesn't get much better.

Monarchs are well known in the three largest and contiguous countries of North America. In Mexico, it is *la mariposa monarca*. In the United States, the monarch butterfly. In French-speaking Canada, *le papillon monarque*. Throughout North America, *Danaus plexippus* shows up regularly through-out its range, depending on the season. In spring, the insects fly north from Mexico, and over the summer they move through the states to southern Canada. As fall approaches, they head back south, ultimately to return to their mountain roosts by November. All along the journey, they arrive as adults, lay eggs on milkweed, and morph through their caterpillar stages to become chrysalises and adults again. The cycle of metamorphosis takes about a month to complete and gives those who take notice ample opportunities to acquaint themselves with the international travelers. It's no wonder monarch butterflies are among the most recognized, studied, and beloved of insects in North America, some say in the world.[1]

Their compelling tale begins when a monarch female deposits a minute, cream-colored egg on the underside of a leaf—anything in the milkweed, or *Asclepias*, genus. Females lay an average of seven hundred eggs, one at a time, on milkweed plants. At first glance, the eggs look like pearly white dots no bigger than the head of a pin. On closer inspection, they reveal themselves as tiny cathedral-like domes, etched with parallel grooves that meet in a subtle tip, pointing skyward.

A few days later, the egg turns dark. Shortly thereafter, a tiny, black-headed, white-bodied caterpillar emerges. Its own eggshell serves as a first meal. Until the baby caterpillar eats milkweed and absorbs the plant's complex chemicals, which are later manifested in the monarch's bold colors and bad taste to predators, its appearance is nondescript.

Monarch caterpillars have adapted to absorb the toxins in the milkweed leaves, the organic substances called cardiac glycosides, which their bodies process in a way that doesn't hurt them. This process is called sequestering, and it's extremely useful in a Darwinian sense. Storing the chemicals without harm makes the butterflies taste nasty to predators. Most birds will eat a monarch butterfly only once. After that first bite, they understand that monarchs taste bad and leave them alone. The butterflies' bright orange coloration warns predators to stay away.

After hatching from its egg, the caterpillar commences a ten- to fourteen-day eating binge, consuming enough milkweed to grow to almost two thousand times its original body mass before becoming a butterfly.[2] During this feeding frenzy, the caterpillars do nothing but eat, poop, and grow. They sport a series of distinctive, ever larger, yellow, tan and black pinstriped suits. In the course of their caterpillar cycle, they will outgrow, shed, and replace this attire five times, often consuming the skin they shed as a meal.

Finally, the caterpillar is ready to make its chrysalis. In the fifth instar—or final stage—the caterpillar gets so turgidly plump it looks like it will bust its stripes. That's when it stops eating. The caterpillar may wander away from the milkweed plant to find a safe place out of direct light to take on its new form. This could be on a nearby twig or shoot, under a patio chair or coffee table, on the lip of a window or the side of a potted plant. Caterpillars raised in captivity have been known to form their chrysalises in the most unlikely places—on curtains, under chairs, even on electrical cords and door jambs. In San Antonio, a city-owned milkweed patch included a shaded, concrete patio overlooking a stand of milkweeds along the San Antonio River. The site often hosted weddings and celebrations and served as a breeding ground for monarch butterflies. The cement deck's floor jutted slightly into the milkweed patch, providing an overhang of several inches. Spent monarch chrysalises in various stages were frequently found in its shade, the empty paper-thin shells abandoned by adult butterflies that had taken flight from the milkweed patch.

Once a location is chosen to form the chrysalis, the caterpillar sits quietly for about a day, seeming to ponder the possibilities. In apparent meditation, it no longer feeds but instead spins a tough, sturdy silk button that can support its entire body weight. After spinning the silk, the caterpillar assumes a J-shape, hanging vertically, head down, from the silk tab for as long as half a day. Then, at some moment, the caterpillar sheds its stripes and forms a jade green chrysalis. The event takes all of five minutes.

The chrysalis hangs, suspended by its self-sown silk button, for ten to fourteen days, depending on the weather and humidity. Its gold dots reflect and deflect sunlight, sometimes causing a bit of a mirage, similar to dew drops. When the time arrives to become a butterfly, the green fades, then turns dark, then black, then clear. Gorgeous orange-and-black wings peek through the transparent shell.

With no fanfare or warning, the paper-thin chrysalis shell splits. A mound of orange-and-black life falls out. First comes the head—lush, velvety, and black. Then a soft jumble of wings, legs, and antennae. The butterfly is extremely vulnerable at this stage. If it drops to the ground and its wings harden in a crumpled fashion, it is crippled for life; since it can't yet fly, a spider could take it for supper. The butterfly hangs helpless, clinging only to the diaphanous chrysalis shell or perhaps a nearby twig or leaf for a spell, until its wings drop completely, take shape, and form properly. After about two hours, the butterfly is ready for its first flight. It opens and closes its wings slowly, repeatedly, seeming to rev its engines. After a month as egg, caterpillar, and chrysalis, the capricious moment arrives: the butterfly takes flight.

The story of a butterfly's birth never ceases to fascinate newcomers. But that's just the first date. Monarchs are incessantly interesting. Their bold, stained-glass wings stand out and repel predators. And yet they don't sting or bite. A dreamy flight pattern suggests confidence, while their elusive flits and turns connote flirtatiousness. Turning legs into wings—now that's magical. And finding their way to the oyamel fir forests of the Mexican mountains, a

A monarch caterpillar makes its J-shape before forming its chrysalis.
Photo by Monika Maeckle.

place they've never been, demonstrates tenacity, strength, and science cloaked in a special kind of magic.

In March, as the spring equinox approaches, millions and millions of monarch butterflies ready themselves for the epic multigenerational migration that begins in the sacred firs of the Trans-Mexican Volcanic Belt mountain range in the states of Michoacán and México. The annual rite is unique in nature. Birds leave from and return to the same habitat in spring and fall, often returning to the exact nest where they hatched their offspring; whale mothers personally introduce their young to familiar feeding grounds where food is plentiful. But monarch butterflies are special, and this explains their unique charm. The creatures that leave Mexico's high-altitude forests in the spring will never return. In early spring, they rouse from a winter of reproductive diapause—a state of suspended sexual activity caused by physical and environmental cues. A butterfly orgy takes place. Males pounce on females when they leave their roosting clusters to puddle for a drink of water.

Male monarchs have tiny, pincer-like claws on their abdomens. They're famous for using these wicked grabbers to overtake females. They unceremoniously swoop down on the ladies in flight, wrestling them to the ground and forcing their spermatophores—capsules packed with sperm, nutrients, and the protective chemicals that make monarchs taste bad—literally inside the females. A male grabs the female by the abdomen with his pincers and locks on, often whisking her away on a flight of "forced copulation."

Scientist Miriam Rothschilde famously accused male monarch butterflies of butterfly rape, citing them as "nature's prime example of the male chauvinist pig."[3] The female may resist and escape, but if the male prevails the couple can be bound for hours in what is euphemistically called a "courtship flight."

Males are visibly assertive in guaranteeing their butterfly DNA makes it into the gene pool to produce the next generation. Male monarch butterflies, like other lepidopterans, have been documented lingering near mature chrysalises. As soon as the newly hatched female's wings have dropped, dried, and assumed flight readiness, the male descends on her. He grabs her with his pincers and has his way with her, bestowing the virgin monarch with his "nuptial gift."

Monarch butterflies copulate multiple times with multiple partners, especially at the overwintering sites where the clock is ticking on the males' chance to reproduce. The sense of urgency to procreate is extreme. A female can literally burst from excess spermatophores inserted in her body. The multiple sperm packets can make up to 10 percent of a female monarch's body weight, overwhelming her and causing death.

After promiscuous mating, the butterflies start their journey north, the females laden with fertilized eggs that will launch the next generation. They head for the U.S.–Mexican border, and by mid-March, just as the antelope horns milkweed is pushing its first tender shoots from the caliche soils of the Texas Hill Country, the monarch butterflies start to arrive in ones and twos in South and Central Texas.

They're looking for milkweed, the monarchs' singular host plant, and the only plant on which they will lay their eggs. Without milkweed, monarchs would cease to exist. The females, often tattered with faded wings, hover over the plants, tuck in their hind ends, and deposit a single egg on the underside of a milkweed leaf. Some butterfly species lay their eggs in clutches, or batches. But female monarchs lay their hundreds of eggs one at a time. Soon thereafter, they die. The males expire shortly after their courtship flights.

Once the female lays her eggs on a milkweed leaf, the egg moves through its month-long cycle—caterpillar, chrysalis, adult. In late spring, the first generation of monarch butterflies hatches. These first-borns continue the cycle, mate, search for host plants, and follow the milkweeds north. By late spring and into the summer, monarch butterflies east of the Rocky Mountains arrive in their primary breeding grounds in the midwestern United States. This is where the migrating monarch butterfly population multiplies. A smaller, separate population of western monarch butterflies migrates up and down the Pacific coast and roosts mostly on the California coast. Western monarchs occasionally find their way to the Mexican roosting sites. The cycle continues. By the summer solstice in mid-June with days at their longest, monarchs are in prime reproduction mode.

They continue this rhythm, what Cornell University ecologist Anurag Agrawal calls "waiting, mating and migrating," through the summer.[4] They arrive in southern Canada, the farthest reach of their range, in late August and early September. Then, around the fall equinox in mid-September, millions and millions of monarch butterflies east of the Rocky Mountains make a big life cycle change. From Toronto to Texas, most stop having sex and head to Mexico.

During the day, they fly south in what's called "directional flight." They stop for fuel along the way, building up their fat stores from sips of nectar to make it through five long celibate months of winter in the Mexican mountains. Late-season nectar plants become more important than milkweed during this stage of the monarchs' life cycle. To conserve energy, the butterflies catch thermal winds and air currents to expedite their travels, sometimes traveling in small groups to "draft" each other like cyclists to reduce wind resistance. At night they drop from the sky, find a welcoming tree, preferably near water and nectar, and form overnight colonies with fellow travelers like those in Hext or along the Llano River.

Roosts and group travel become more common as the butterflies migrate south, congregating in the Texas Funnel in October, late in the migration season. As the autumn sun warms to at least 60 degrees each day, the monarchs resume their travels, averaging forty miles per day. In late October and early November, they arrive at their destination—the Trans-Mexican Volcanic Belt, a high-elevation mountain range that straddles central Mexico.

The fact that no single butterfly makes this epic round-trip is unique in nature. The monarch butterfly that arrives in Michoacán around Día de los

Muertos (Day of the Dead) in early November has never pumped its wings anywhere near this part of the world. The completion of a three-thousand-mile journey with no guidance from elders or familiarity with their destination contributes to our fascination with the species.

It also remains a scientific puzzle. How do they find their way? Using their sun compass and internal body clock? Magnetic fields? Following a pheromone trail left by those that preceded them on the journey? This mystery has intrigued scientists and enthusiasts for decades and motivated the "discovery" of the monarch butterfly roosting sites in 1975.

CHAPTER 3

CATALINA'S STORY

Catalina Aguado Brugger and her husband Ken retired early to their motor home in Zitácuaro, Michoacán, on New Year's Day, 1975. They were not the type to overdo it on Mexican fireworks or tequila. The Bruggers were simply exhausted from strenuous ten-mile hikes over the previous few days, hikes that climbed to elevations approaching 11,000 feet.

The Bruggers had been combing the mountain trails surrounding their base in the small town of Zitácuaro for the better part of a week. They were trying to reach Cerro Pelón, a naked geological formation that sits at 9,474 feet and translates to "bald hill" in English. The couple had spent many weekends over the past two years exploring the remote, thinly populated mountains west of Mexico City, traversing the roads and forest trails of the Trans-Mexican Volcanic Belt. Their goal: to locate the ancestral roosting sites of millions of monarch butterflies that migrate each year over multiple generations from Mexico, through the United States, to Canada, and back. If the Bruggers accomplished their objective, they would solve a mystery that had dogged scientists for decades.

Ken, a smart, charming *norteamericano*, first became intrigued by an ad in the *Mexico City News*, an English-language newspaper that catered to expatriates. According to the newspaper, Canadian scientist Fred Urquhart was looking for volunteer "research associates" to help track monarch butterflies. Urquhart and his wife Nora had been studying monarch butterflies since the 1940s, trying to determine where the migratory insects spent their winters. They now believed the butterflies were overwintering somewhere in Mexico, and they needed local volunteers to help them prove it.

"I read with interest your article on the monarch," Ken wrote to the Urquharts in February 1973. "It occurred to me that I might be of some help."[1] The

Urquharts accepted Ken Brugger's offer of assistance. The job was voluntary at first, and Ken thought it would be fun. "C'mon, want to do it?" he asked Catalina, his newlywed wife. At first she hesitated. "Good luck with the campesinos and the Mexican government," she said.[2]

Ultimately, both Bruggers joined the effort, and now here they were, this brisk winter morning at the trailer park in Zitácuaro. The day started early with cinnamon sugar on cream of wheat, a hot cup of Sanka, and condensed milk. Their plan today was to try, once again, to reach Cerro Pelón. Catalina and Ken knew the butterflies were up there. Not long ago, when Ken was driving solo through the mountains west of Mexico City, he'd encountered a slew of dead monarchs in the road that had been pelted from the sky by hail. He and Catalina had also seen groups of the fluttering masses in their travels in recent weeks.

Despite such encouraging signs and an early start each of the previous four mornings, the couple's search had ended in failure. It didn't help that winter sunsets came early in the chill mountain afternoons, forcing a return to camp before nightfall.

Today would be different. For the first time, someone from the village would join the search. It was Catalina's idea. She thought it wise to recruit a local, someone familiar with the trail.

As a native Michoacána, Catalina sensed the suspicious gazes cast upon the couple as they traveled through the area. This was not a place accustomed to tourists. Locals didn't always accept that she and her husband were looking for butterflies. The *chisme*, or local gossip, likely included speculation that they were drug dealers, smugglers, sex traffickers, or worse. And granted, fifty-seven-year-old Ken, a self-taught engineer who worked at textile giant Rinbros in Mexico City, and the vivacious, twenty-five-year-old Catalina, a native of the nearby colonial city of Morelia, struck a provocative and unconventional profile as they careened through the remote mountain villages on Ken's motor scooter. Ken, from Kenosha, Wisconsin, had moved south in 1965, after a divorce. Catalina, who wanted anything but to be a married teenager like her three sisters, left her family's ranch outside Morelia for Mexico City to travel and work in sales. A friend introduced them.

The couple had now been traversing the mountains of Macho de Agua, El Capulín, Popocatépetl, and the Nevado de Toluca for almost two years. They slept at local inns or camped in Brugger's twenty-foot Winnebago. Rather than build campfires that would call attention, they heated up cans of beans on the warmed engine block of their vehicle.

"Have you seen these butterflies?" Catalina would ask the locals when they stopped along the way, sharing a mounted butterfly and photos of monarchs in various stages—caterpillar, chrysalis, adult. Urquhart had sent the samples from his collection in Canada to assist with the search. "We're doing it for science," Catalina explained. She must have asked hundreds of people over the two years they searched. Not one person ever admitted to having seen the butterflies, perhaps an instinctive act of protection of a unique Mexican treasure.[3]

As Catalina and Ken descended from their daily hike on January 1, Catalina encountered an elder named Agapito along the trail with a horse. She asked him to join the hunt. "Do you think you could come with us tomorrow?" she asked in her soft-spoken Spanish. "We are looking for these butterflies." The aging campesino stood alongside a horse and two scruffy dogs, mulling her proposal.

"We pay you for the day," she told him. "We need to take the horse because we need to take food and cameras."

Agapito seemed interested, but hesitant. The horse belonged to his uncle.

"Do you want to ask him if he can lend it to you? We'll pay you double if you take the horse," Catalina said. "We'll wait for you to ask him."

Agapito disappeared for a bit, then returned. "Can we meet you in the morning?" Catalina asked. They agreed to meet.

Catalina and Ken awoke the next day at 3 A.M. They drove to the trailhead to meet Agapito. With Cerro Pelón hidden among towering oyamel trees, they loaded their gear on the horse. Up they went.

▼

Catalina had always displayed a sense of curiosity and adventure. As a child, she snuck off to the library at the Universidad de San Nicólas de Hidalgo to peruse books on science. She gardened at her family's ranch. And while many young Mexicanas might respond with trepidation to the exotic insect life encountered in the great outdoors, Catalina found invertebrates enchanting. Especially the butterflies. After a good mountain rain, she loved to watch

Mexican bluewings, vibrant fritillaries, and exotic swallowtails puddle in a seasonal stream near her home.

At age twelve, Catalina moved with one of her sisters from the ranch to Morelia. By seventeen she was living in Mexico City, working at a pharmacy, and later in sales for Philips Comercial. She loved to travel and roamed the hemisphere as a fearless, free spirit, exploring Mexico, Guatemala, El Salvador, and elsewhere. She traversed the United States and Canada, alone and with friends. They boarded buses and slept in cheap hotels, sometimes camping along the way, sating their wanderlust and search for adventure.

Catalina was a well-traveled twenty-one-year-old when she met Kenneth Brugger. A Canadian friend introduced them in Acapulco. She was taking a break from work, visiting the Pacific Coast resort to attend immersion English classes.

Naturally, the fifty-three-year-old Brugger found Catalina captivating. Romance between older gringos and younger women is not unusual in Central America. Expatriate gringos, many retired and known as *pensionados*, often marry young Latin American women. Some countries, including Costa Rica, have even created special immigration status for these elderly husbands, who enjoy tax breaks and duty-free imports.

Mexico did not have an official pensionado program, but its low cost of living, proximity to the United States, and welcoming weather convinced Brugger to move there after his divorce. A good man with a steady income, a motor coach, and a scooter, he offered Catalina the appealing prospect of both security and adventure. By her account, he was also a charmer. After they met, he followed her around. When she left for a trip to El Salvador, he wrote her sweet notes on the back of a tortilla.

They married in 1973. Then Ken saw the Urquharts' article in the *Mexico City News* seeking help in tracking down the monarch butterflies' winter home. He wanted to participate and volunteered in a letter, typed by Catalina. Fred Urquhart accepted the offer.

In the winter of 1975, the couple ended up in Morelia, tracking the monarchs. Meanwhile, Urquhart had received reports that at least some of his experimental monarch butterfly tags had been recovered northwest of the capital. In October 1974, the Bruggers saw pulses of monarchs moving west from Mexico City, suggesting the butterflies were heading to Michoacán.

"You must be getting really close," Urquhart wrote them in an encouraging letter. After almost a year of voluntary progress reports to Urquhart, the Bruggers received their first financial compensation for their time—about $3,000, Catalina recalled. Enough to cover travel, boots, topographical maps, and other supplies. That's when they started taking it more seriously, she said.

▼

With their local guide, the trek was much quicker than previous attempts. Agapito knew a shortcut. A horse to haul their gear expedited the trip. The threesome moved up the mountain, climbing more than 3,000 feet in altitude on rugged trails in several hours. As they wove through Montezuma pines and grand stands of oyamel, or sacred firs, Catalina noticed Agapito growing annoyed. He seemed angry, short, even disgusted. He confessed a fear of the coral snakes that inhabit the area, and the wrath of his uncle if one were to bite his horse. Catalina assured him everything would be fine.

Using a pair of wooden walking sticks, one in each hand, Catalina waded through the leaf litter, moss, and rocks. She wore a light, broad-brimmed hat, turtleneck, motorcycle jacket, and hiking boots. At one point she dipped into a hole, then recovered, practically walking on her knees, because of the soft, mossy ground. She worried, since that's where the snakes hide.

The annoyed Agapito led the horse as they continued the climb, approaching the top of the bald hill, stepping lightly.

Catalina was first to the roosting site. "I see them! I see them!" she yelled, taking in the view at Cerro Pelón. It was chilly and the butterflies clustered in dense masses to stay warm. Their tawny-gold underwings created a thick, lush blanket that covered the tree trunks and limbs. With the air still and the butterflies motionless, a lone butterfly would occasionally leave the group and drift in search of a more welcoming spot.

Catalina and Agapito stood immobile, taking it all in, waiting for Ken to catch up. They removed the tripod from the horse and set up the camera in the clearing to face the forest. Using an old-school, wired remote with a thirty-second timer, Catalina snapped a photo that appears to be the first selfie ever taken at the monarch butterfly roosting sites. In the photo, she gazes at the massive trees covered with butterflies. Her posture assumes what appears a triumphant stance, a walking stick in each hand, her right leg cocked slightly uphill in support of the left, projecting We did it! Agapito and the horse can be seen in the background.

Over the following days, Catalina and Ken found two other monarch butterfly colonies, both equally astounding as the first. On January 9, 1975, they placed a call to Urquhart from Tuxpan to relay the news. "We have located the colony!" Ken told him. "We have found them—millions of monarchs—in evergreens beside a mountain clearing."

Twenty months later, Catalina graced the August 1976 cover of *National Geographic*. The now iconic cover photo has her sitting on a stump, completely engulfed by monarch butterflies. They dance on her blue shirt and jeans, alight on her head, and rest on her arms—which she extends in a joyous reach. Delight spreads across her tanned face. "Discovered: The Monarchs Mexican Haven," read the headline to a story written by Fred Urquhart himself. The story rocked the world of lepidoptery and launched a generational rivalry between two scientists that would shape monarch butterfly science for decades to come.

▼ ▼ ▼

CHAPTER 4

EGO WARS
An Unpleasant Incident
in the Forest

Just months after the publication of the *National Geographic* cover story, Ivy League–educated entomologist Lincoln Brower stood in the forest of Sierra Chincua, Michoacán. The location, at 11,500 feet altitude, sat about two hours north of the site at which Catalina and Ken first walked the roosting sites two years before. Brower wore a turtleneck sweater and dark corduroy trousers, a canteen draped casually across his back. Fred Urquhart crouched on the forest floor. He sported a leather jacket and knit cap to ward off the chill mountain air. Boxes, walking sticks, a tarp, and other supplies were strewn in the mulch of the forest floor around him. A butterfly net lay in the foreground. As Brower towered over Urquhart, his left hand propped assertively on his hip, he looked down on the famous professor twenty years his senior. Urquhart lifted his head. He glared at Lincoln Brower with an exasperated look, as if to say, What the hell are you doing here?

Brower extended his hand to greet Urquhart, but Urquhart refused the gesture. A photograph snapped at the particular moment of their infamous encounter in the mountains of Michoacán perfectly captures the absolute disdain the two scientists had for each other. The tense moment was appropriately labeled an "unpleasant" incident in the forest.[1]

The snub was the latest in a series of rejections that began in 1973 and forever defined the relationship of perhaps the two most important monarch butterfly scientists of the twentieth century. How they ended up on the same mountain that cold winter day has become one of the seminal stories of monarch lore. Separated by a generation, yet having shared many similar life experiences and singular scientific achievements, the two had created a

The kismet moment of the Brower-Urquhart mountaintop encounter,
captured in a faded photograph, became one of the seminal stories
of monarch lore. "As you can see, he's not too happy," said Lincoln
Brower. Photo courtesy of University of Florida Biography.

rift that became insurmountable, as steep and convoluted as the mountain
trails that led them to this awkward moment.[2]

Urquhart, a Canadian zoology professor at the University of Toronto, with
his wife Nora, had realized the most important discovery in the field of butterfly biology of the twentieth century. The couple had answered a question
that vexed butterfly followers for more than a century: where do millions of
monarch butterflies spend the winter?

Urquhart had become fascinated with butterflies, monarchs in particular, as a child. Each August the orange-and-black visitors would arrive
in Toronto for a short stay, then suddenly depart. Where did they go? The
question drove him for decades. After thirty-five years, including a quarter-
century mobilizing thousands of volunteers through his Insect Migration
Studies Association and the dedication of Ken and Catalina, Urquhart had
finally solved the mystery.

Brower, a biology professor at Amherst College and a pioneer in the field
of chemical ecology, had also been smitten with butterflies as a boy. At age
five he participated in family tennis matches in Chatham, New Jersey, where
his family ran a commercial rose farm. When the game became overwhelming, Brower plopped down on the grass for a rest where he noticed a variety

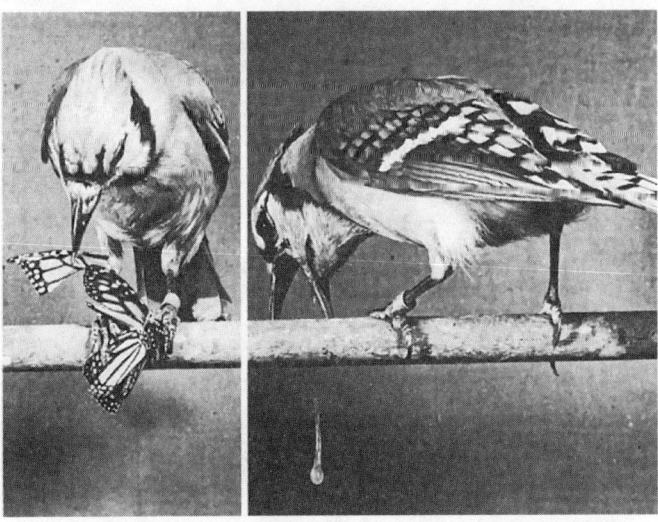

Brower's famous "barfing blue jay" photo of a bird retching after
eating monarchs, proving that monarchs don't taste good.
Photo courtesy Dr. Lincoln Brower.

of insects flitting on weeds and clover. He found *Lycaena hypophleas*, a
thumbnail-sized copper butterfly, especially captivating.[3]

Studies by Brower and his first wife Jane Van Zandt led to the discovery
that monarchs sequester or store toxic chemicals in their bodies that render
them distasteful to predators. Further research led Brower to answer another
much-debated question: do the butterflies produce the toxins themselves,
or do they absorb them from the milkweeds they eat? Brower and Van
Zandt's famous "barfing blue jay" photo, which depicts a blue jay retching
after consuming a monarch butterfly, proved unequivocally that monarch
butterflies are what they eat—toxic and distasteful.[4] Working with other
scientists, Brower went on to develop a chemical fingerprinting methodology
that allowed him to determine the provenance of the milkweeds monarch
butterflies consumed.

Looking at these two famous monarch butterfly biologists, both of whom
discovered the magical creatures as boys, married fellow scientists who
became true partners in their research and adventures, and made significant, historic contributions to butterfly science, it's impossible not to wonder
what would have happened had they worked together. But that was not to be.

After Catalina and Ken Brugger located the roosting colonies in 1975, Fred and Nora Urquhart announced the news in their annual newsletter, which reached the more than three thousand members of their Insect Migration Studies Association. Volume 12 touted 1975 as "a most eventful year" and "the year of Mexico." In it, the Urquharts announced they had located the butterflies' overwintering site but, for the moment, were not yet disclosing its location. Instead, they invited their associates to watch for an upcoming article in the widely read *National Geographic* magazine.

News of the discovery spread quickly. When Brower heard about it, he contacted the Urquharts, with whom he had become acquainted over the years. Brower found Urquhart "really pompous."[5] Their relationship had been professional but cool, with the scientists exchanging letters on and off for years. Brower sent Urquhart a reprint of a study comparing the chemical constitutions of eastern monarchs from different regions. Urquhart reciprocated by sharing some unusual milkweed seeds with Brower.

This was not the first time Brower had asked Urquhart about the roosting sites. Previously, he had asked Urquhart of their whereabouts via letter,[6] but he failed to get an immediate response. Upon reading about the discovery in the newsletter, Brower felt a sense of urgency and pressed Urquhart again. Details of the location would remain secret until the *National Geographic* story was published, Urquhart responded. Frustrated, Brower contacted the magazine directly to see if they would divulge the location. They also refused.

Then, in August 1976, Catalina's dramatic *National Geographic* cover portrait hit newsstands and mailboxes. Rich color photos showed millions and millions of monarchs—flushing in an orange storm, shrouding the towering sacred firs, massing on damp streambeds to puddle for water. Fred Urquhart, sixty-five years old at the time of his first trip to the colonies, wrote vividly of their challenging climb up the 10,000-foot mountain. His heart pounding, his feet leaden, he considered his own mortality at the same moment he was realizing a lifelong dream. "The rather macabre thought occurred to me: Suppose the strain is too much," he wrote. "Then we saw them. Masses of monarch butterflies—everywhere!"

The sensational photos stirred Lincoln Brower to his core. All the migrating monarch butterflies from the eastern United States aggregated in one place? What a fantastic natural laboratory-in-the-wild this would be![7] The roosting sites represented an ideal place to test chemical ecology theories. Were the birds eating all the butterflies or just some? Which ones? Were the butterflies equally toxic and distasteful?

A month after the *National Geographic* story, Urquhart wrote another article—this one for the Lepidopterists' Society newsletter. He repeated much of the same information but for the first time revealed that the butterfly colony was situated on a volcanic mountain in the northern state of Michoacán. This contradicted the title of the article, "The Overwintering Site of the Monarch Butterfly in Southern Mexico," which referred to the other end of the country. Michoacán is considered part of central-western Mexico. Meanwhile, the *National Geographic* article mentioned Mexico's Sierra Madre as the monarchs' spectacular winter hideaway.[8] The geography didn't add up.

In October 1976, Brower contacted Urquhart yet again and asked him to share the location of the colonies. He invited Urquhart to Amherst to present a lecture on the discovery. Brower pressed Urquhart: "Perhaps in view of this letter, you might review your position and consider sharing the location of the site with a fellow scientist, who, like you, is equally keen in conserving the site from modern depredations of human society. Again, I congratulate you upon your discovery."[9]

Urquhart took two months to reply. Perhaps Brower should visit south Florida—Apalachicola Bay, specifically, Urquhart suggested. There, said Urquhart, Brower could gather specimens of migrating monarchs that he claimed to so desperately need. It would accomplish the same thing, Urquhart wrote to Brower.[10]

The dismissive advice infuriated Brower. Though he sympathized with Urquhart's desire to protect the colonies from exploitation, Brower believed scientists had an obligation to collaborate and share information. To him, the misleading geographic references smacked of arrogance and obfuscation—especially in response to a fellow scientist. He labeled Urquhart completely ignorant of Mexican geography and accused the Canadian scientist of deliberately deceiving the world about the location of the butterflies.[11] He found the Sierra Madre reference especially preposterous. "That sounds nice," Brower said years later, "like a Humphrey Bogart movie or some damn thing."

Urquhart wouldn't budge. The secret location of the monarchs' winter home remained safe—for now.

Then, at a butterfly talk in Massachusetts, Brower met an easy-going Texan named William H. Calvert who split his time between Austin and Boston. Calvert studied philosophy and physics as an undergraduate at the University of Texas at Austin. By the time he met Brower, he was studying tent caterpillars, researching how female butterflies found their host plants, in pursuit of a zoology PhD. Calvert had a reputation as a cowboy entomologist resulting from his Texas roots, a preference for fieldwork over the lab, and a distaste for academic bureaucracy.[12]

The two chatted. When Calvert learned Brower was devoting his life to the study of monarch butterflies, he offered to retrieve a couple hundred from a roost he'd come to know in Bustamante, Mexico. Calvert had become acquainted with the area from trips to the karst caves of Nuevo León in northern Mexico with the Austin spelunking club.[13] Brower liked the idea and offered to pay Calvert's expenses. On his next trip home, Calvert drove to Mexico from Austin and returned to Amherst a few weeks later with two hundred monarch butterfly specimens stored in glassine envelopes in an ice chest. Brower ground the butterflies into a paste to examine their chemical makeup, and a fluid, long-term collaboration was born.

Fed up with Urquhart's snubs, Brower and Calvert hatched a plan: Calvert would lead an expedition to Mexico to locate the roosting sites, and Brower would fund it with grant money from the National Science Foundation. They carefully reviewed Urquhart's written accounts and realized that "Uncle Fred," as Brower sometimes called Urquhart,[14] had dropped two important clues regarding the location of the roosting sites. He mentioned that the volcanic mountains sat somewhere around 10,000 feet in altitude. And he noted that they were located in northern Michoacán province.

Based on those clues and consulting relatively detailed topographical maps, Brower and Calvert determined that a small area west of Mexico City looked promising as the location of the roosting sites. On December 26, 1976, Calvert and John Christian, a Spanish-speaking associate from the University of Texas, set out in a beat-up cream-colored Ford pickup truck. They drove to Morelia to pick up two Spanish-speaking friends: Victoria Foe, a fellow biologist and future McArthur Fellow, and her boyfriend Michael Dennis. All four squeezed into the single bench of the pickup. They set out for Angangueo, a small lumber and mining town about seventy-five miles west of Mexico

City in the state of Michoacán. Tucked into the Trans-Mexican Volcanic Belt at an elevation of 8,500 feet, Angangueo was one of few locations that fit the criteria Brower and Calvert had discerned from Urquhart's writings. It seemed as good a place as any to launch their search.

Upon their arrival in Angangueo, Christian, who grew up in Mexico and spent years researching Indigenous peoples there, did most of the talking. He asked around town for the mayor so they could request written permission to hike the mountains. This is a suggested protocol in parts of Mexico governed by *ejidos*, communally owned lands awarded to Indigenous peoples after the Mexican revolution in the 1920s. Upon locating the mayor, Christian and Calvert pulled out a monarch butterfly pressed in a plastic sheath as an example of what they were seeking. Angangueo mayor Manuel Arriaga Nava recognized the creature and seemed flattered and incredulous that anyone would have any interest in the masses of orange-and-black insects that occupied their local mountaintop each winter.[15] He provided written authorization for them to ascend the mountain and assigned his nephew Alian to guide them on the rugged trails.

The group set out the following morning, Thursday, December 30. It became quickly obvious that Alian was not all that familiar with the location of the colonies. He led them on an exhausting, serpentine hike, up and down the rugged trails, ascending, descending, and backtracking all over the mountain. Finally, in late afternoon, they arrived at the roosting colony at Sierra Chincua. The butterflies were aggregating for the evening, gathering in masses and fluttering to rest on the warm, robust trunks of the towering oyamel firs. The sun was just beginning to set along the western wall of the Sierra, and its warm, backlit glow cast an astonishing sight.

It was getting late, and the group still faced a strenuous descent before nightfall. They started down the mountain with little fanfare. Everyone was exhausted. They spent the night in Zitácuaro, more than an hour from their hike's starting point. Calvert was suffering from altitude sickness, and the slightly lower altitude helped assuage his headache and nausea.

The next day, New Year's Eve, Lincoln Brower was preparing for a rousing celebration at his home in Amherst, Massachusetts. About an hour before 1977 arrived, Calvert called from Mexico. The connection was awful, but the news was excellent: they had found the butterflies, almost two years to the day after Ken and Catalina chanced upon the colonies at Cerro Pelón. The Calvert team accomplished in four days what it took Urquhart almost four decades to achieve. It was a very happy New Year for Lincoln Brower.

About three weeks later, Brower, Calvert, and a research team returned to the site. They flew to Mexico City, rented a pair of Volkswagen vans, and drove to Angangueo. Alian, from the village, joined them again. They navigated the bone-shaking dirt road up the mountain as far as possible by car, parked, and unloaded their gear for the demanding hike. Up to 11,000 feet, back down to about 9,500, Brower recalled that the group almost didn't make it.[16]

About three in the afternoon, Brower and crew arrived at Urquhart's base, and the rejected handshake episode unfolded. "It was not a pleasant meeting, to use an understatement," Brower told the *New York Times* months later.[17] Science, it seemed, was taking a back seat to Urquhart's sense of ownership in what appeared to be an old-fashioned turf war.

Calvert had to return to Boston for work and could only stay for the weekend. Brower and the others remained to conduct research for two weeks. At some point in his myriad entries and exits from the forest, Calvert dropped his field notebook along the trail. Catalina and Ken were assisting the Urquharts with their research, and Ken found Calvert's notebook. Penned inside its cover: Calvert's home address in the Texas capital.

Brugger shared the notebook with Urquhart, and Calvert's Austin address became the basis for a rumor in which Urquhart deduced that Calvert had followed him all the way from Austin to Sierra Chincua. This would have been a remarkable feat, if true. It wasn't. Nor did Calvert ever recover his notebook.

Another misunderstanding resulted when Alian responded to the cold by building a small campfire, a common practice in the forest. Smoke from the fire wafted over to the roosting butterflies and caused some to drop to the ground. Urquhart viewed the episode as an affront and accused Brower of starting a forest fire to flush them from the trees to create dramatic footage for the movie he was recording.

Antagonism mounted from these misunderstandings, and the standoff of the butterfly biologists made news, resulting in a public relations war between the two men. Urquhart fueled the story by sending a "special report" to his thousands of Insect Migration Studies Association members. Alian's innocent fire in a part of the world where people warm their homes with fireplaces and wood-burning stoves was portrayed as an uncouth and destructive act. Urquhart cast Brower as an outlaw who deliberately set fire to the butterflies' house to create narcissistic drama for his movie.

"He misinterpreted things all along the way," Brower said years later.[18] In late 1977, Brower countered Urquhart with his own version of events in the Lepidopterists' Society newsletter. He set the tone in the first sentence, establishing that for purposes of the discussion Sierra Chincua would be known as "Site Alpha."

The feud continued for decades. Urquhart never accepted the science that monarchs are unpalatable to predators because of their consumption of toxic milkweed leaves. At more than one public talk, he ate a monarch butterfly and announced that it tasted like dry toast.[19] In his 1985 book *The Monarch Butterfly, International Traveler* he completely dismissed Brower's chemical ecology research: "By the use of abstruse terminology, the research assumes an aura of highly qualified investigations, but, when carefully analyzed, contains nothing of real value, and no meaningful conclusions."[20] He frequently referenced examples of birds and cows eating monarchs and discounted the Browers' barfing blue jay experiment, contending that blue jays typically eat seeds rather than insects.

Brower viewed Urquhart as petty for not sharing the sites' location with a fellow scientist. At one point, he labeled Urquhart's science as quaint, sloppy, an "amateurish, self-serving approach to biology."[21]

Eventually Urquhart admitted to obfuscation motivated by good intentions. "We lied," he told *People* magazine in 1978.[22] "We wanted to keep the place secret until it was protected." In the same interview, he vowed never to return to the forest where he first found the butterflies and suggested that other yet-to-be-identified colonies awaited discovery. In the future, he said, "We will go silently and unobtrusively." And as for Lincoln Brower, "I leave him with his studies of poisons. I have my studies of migration."

Neither scientist could have anticipated that the public would be as captivated by monarch butterflies as they themselves had been, a collective fascination that developed into what became known as citizen science.

▼ ▼ ▼

CHAPTER 5

CITIZEN SCIENCE
From Alar Tags to iPhone Apps

Fred and Norah Urquhart had few options for promoting their studies of the monarch butterfly migration in the 1950s. But like many people before and since, the scientists used the media as a strategic tool. An article placed in *Natural History* magazine helped the Urquharts spread the word of their studies and, for the first time, provided a national platform for recruiting volunteers to help solve the mystery of the butterflies' journey.

At the time, *Natural History* was sent free of charge to all members of the American Museum of Natural History in New York City, which published the periodical. Since the magazine's founding in the early 1900s, its mission had always been to promote understanding of nature and science.

The May 1952 edition included a four-page spread titled "Marked Monarchs," bylined F. A. Urquhart but written by Norah. In addition to her sociology and biology degrees, Norah had an apparent talent for public relations. She wrote and placed many articles on studies by the Urquharts over the years. "Marked Monarchs" appears to be her first media placement in a national publication.

Perhaps Norah had a better way with words and the couple assumed—correctly—that the article would carry more weight if a PhD appeared with the byline. Ghostwriting is common in many professions, and women often took a back seat to men in the world of science at that time.

Their piece in *Natural History* introduced the intrigue of the monarchs' journey to readers of a mass medium for the first time. Urquhart cast the butterflies as "mysterious travelers" that embarked on a multicountry migration, from Canada through the United States, down to Texas and unknown points south. They portrayed the monarch migration as a massive puzzle.

Where did they go? Did they fly independently or in groups? Did they return to their breeding grounds?

The article marked a creative approach by the Urquharts to tap untrained volunteers to observe, collect, and report data back to them at the University of Toronto. The invitation to help unravel the mystery of the monarch migration constituted a 1950s equivalent of citizen science, and with their shared interests in nature and discovery the tens of thousands of *Natural History* readers made a promising target market for participation.[1]

After a description of the monarch's life cycle and a recap of the challenges posed by developing effective tags for data collection, Nora Urquhart closed the article with a plea: "It is hoped that some of the readers of this article will want to assist in the tagging," she wrote, explaining that thousands of specimens would be required from all over the continent to gain an understanding of this unique natural phenomenon. Those interested in becoming "butterfly banders" should send a letter to Toronto and request "tags," specialized stickers for tracking the butterflies. Bird banding had been going on for centuries, but tagging butterflies and tracking their movement had yet to enter the research arsenal—until now.

Fred Urquhart and others struggled for decades to perfect a gummed label that would stick and stay on a monarch butterfly's wing. Much of the challenge resulted from the irregular surface on which the tag must adhere—a tiny patch of thousands of butterfly scales that create an unwelcoming surface for most adhesives. The glue and tag stock required durability in wind, rain, and wild temperature fluctuations over many months. The tag must also include a serial number and instructions for returning it and the specimen, if found, to scientists. It had to be safe for the butterflies and not inhibit natural flight. And, ideally, the tag would be relatively easy to apply.

Combinations of paper, inks, and adhesives (some of which turned out to be toxic) preceded the alar, or wing tags, of today, which slip onto a monarch wing with the flick of a fingernail and light pressure. Experimental tracking approaches included that of C. A. Anderson, of Dallas, who "brands the butterflies with a rubber stamp" and special type of ink. Each insect receives a combo single letter/two digit serial number—"for example Y 32," Urquhart wrote, before providing Anderson's address for reporting purposes, in the event of any recovered and stamped butterflies.

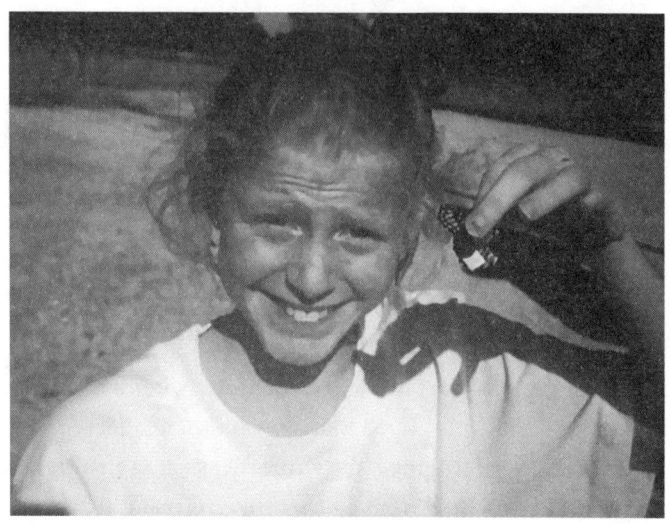

Kati Singleton tags a monarch in 1994 using the glue-and-paper
method. Photo courtesy Jennie Singleton.

Fred Urquhart experimented with dyes and oil paint, applying the latter
with a spray gun. Then he tried a version of the ink stamp. Later he enlisted
the help of the University of Toronto Press to print sheets of fifty tags that
advised, "Return to Museum, Toronto, Canada." This tag had to be applied
with glue, which was cumbersome, water-soluble, and short lived. Eventually
he utilized adhesives that had to be moistened to stick, as with postage stamps.
But upon testing, those failed, too. Finally, the team in Toronto figured out "a
most successful method" for tagging the migrants, which Norah Urquhart
described in the article.

The half-inch by quarter-inch rectangular tags, preprinted with a message
that included a return address and a serial number noted in pencil, were folded
in half and applied with glue to the butterfly's outer wing. Its success relied
entirely on the tag's application, which seems barbaric today, given the ubiq-
uity of self-adhesive stickers on everything from fresh fruit to windshields.

"Punch a hole in the wing of the butterfly with a paper punch," wrote
Norah Urquhart, describing the breakthrough. "Over this hole, which was
made immediately behind the stout marginal vein of the wing, place a tiny
white paper label." She then detailed how the gummed label folded over the
outer edge of the butterfly's forewing and glued onto itself through the hole.

Presumably the weight of the label partly compensated for the missing piece of wing, thus maintaining the butterfly's balance and flight capabilities. The Urquharts calculated that it took eight seconds to apply such a tag—not bad if tagging hundreds or even thousands of butterflies.

Butterfly banders in the Toronto area, organized by the Urquharts, tagged about three thousand monarchs using the Urquharts' method. Of those, only seven were returned to the Museum in Toronto. Those long odds suggested the enormity of the task at hand and led to Norah's plea. Her *Natural History* story resulted in twelve recruits that first year for the Urquhart tagging team. The initial dozen grew to thousands of "research associates" and evolved into the Urquharts' Insect Migration Studies Association, a far-flung group of volunteers situated throughout the hemisphere and loosely bound by their love for butterflies and interest in service. These early recruits became the founding monarch taggers we now know as citizen scientists. Many tagged with the Urquharts for decades, including Don Davis of Montreal, who started in 1968 and continues to contribute as a citizen scientist to this day.[2]

By the time the roosting sites were finally located in 1975, nearly twenty-five years after the Urquharts' initial call to action, hundreds of thousands of butterflies had been tagged by volunteers in almost every state in the continental United States, three Canadian provinces, Mexico, Australia, and New Zealand. By 1987 the Insect Migration Studies Association boasted four thousand associates. Volunteers also reported their sightings and observations, and sometimes sent live butterflies to the Urquharts.

In 1964 the Urquharts began publishing the annual *Insect Migration Studies Newsletter*. The hard-copy bulletin, typed in Courier font and sometimes exceeding thirty pages, was mailed to all associates. Typically it included a recap of the monarch season, a report on how many butterflies were tagged and recovered, a list of select media placements, and a chronicle of milestone events like the discovery of the roosting sites in 1975. The Urquharts often acknowledged their volunteers with shout-outs and details of unusual escapades—like the time Marcia Chambers of Tulelake, California, chased monarchs on horseback to evade a den of rattlesnakes.[3] "Even under the circumstances, Marcia was able to tag 142 monarchs. Congratulations!"

By the time Fred and Norah Urquhart, ages eighty-three and seventy-four, respectively, announced their decision to cease its publication in 1994, the *Insect Migration Studies Newsletter* had been assembled and distributed for thirty years. The couple had spent more than five decades researching, chasing, and sharing their understanding of monarch butterflies.

"Fifty seven years have passed since the first alar tag was placed on the wing of a migrating monarch butterfly in 1937," the Urquharts wrote in their 1994 annual report to research associates. "We have answered questions not only about the migration but also numerous biological questions respecting morphology, development and factors involved in the life cycle."[4]

▼

The Urquharts' tagging program was well under way by the time Rocio Treviño moved to Saltillo, Mexico, in 1978. Treviño had been in the desert capital for about six months when she first noticed the monarch butterflies roosting in the trees at the Universidad Autonóma Agraria Antonio Narro (UAAAN), the local university where she and her husband worked. Treviño and her family had relocated from their home in Chihuahua, about 350 miles to the northwest. Treviño's husband, Julio Carrera López, taught wildlife, ecology, and natural resource management at the UAAAN. Treviño taught biology there.

The family had settled into their new home when in September, for the first time, Treviño noticed masses of monarch butterflies roosting in the pine and walnut trees on the university's campus. Hundreds, sometimes thousands, of the orange-and-black butterflies rested in the evergreen limbs.

Treviño admits that back then she, like most people, paid little attention to the monarch butterfly migration. A longtime nature lover, beekeeper, gardener, and rabbit and dog breeder, she had heard about the migration when the roosting sites were "discovered" in Mexico, but she hadn't paid much attention. What caught her eye and made her angry, however, was the hunting of the butterflies in the streets of Saltillo.

Local children used tree branches, paper bags, tennis rackets—whatever was available—to knock the butterflies from the trees. The "weapon" depended on the socioeconomic status of the children, she recalled. Some butterflies died, others flew off; sometimes the kids kept them or left them in the street.

Treviño attributed the behavior to ignorance. Local people sometimes referred to the insects as *enchiladas* because of their *chile* orange colors.

They also called them *palomitas*—little doves.[5] *Palomitas* can also mean "popcorn" in Spanish, and, interestingly, the monarchs arrived just as the corn crop was ready for harvest.

Whenever she noticed the kids' rabble-rousing, Treviño would stop and explain where the butterflies came from, and where they were going. If she was driving by, she would lean on the horn and signal the children to stop the harassment.

It took several years for Treviño to get her hands on the *National Geographic* cover story that explained the monarch butterfly migration. She was thrilled to find the 1976 edition among a pile of discarded periodicals stacked on the street. "It was a curious thing. . . . between them, out came the magazine with the cover of Cathy Brugger and the monarchs," she said. When her husband traveled to Mexico City as director of the National System of Protected Areas, he would often bring home educational materials from the office that increased her understanding.

Treviño came to realize that Saltillo lay directly in the butterflies' migratory path. The seasonal visitors arrived like clockwork in early September, stopping to nectar on abundant sunflowers, sages, and lantanas that grew in the semidesert landscape immediately after the rainy season. In the evening, the butterflies roosted in the trees on the UAAAN campus.

Treviño had always been interested in the outdoors and worked with her husband as a volunteer for PROFAUNA, a conservation organization dedicated to protecting wildlife and natural resources and founded by the couple and their colleagues at the UAAAN. In the nonprofit's early days, she worked to raise awareness of the plight of the Mexican wolf and golden eagle. PROFAUNA would stage workshops and informational sessions for teachers and others interested in conservation. The organization functioned as an ad hoc group for many years prior to its official formation in 1988.

When Treviño learned about the monarch migration, she and like-minded colleagues began taking groups of students who visited the university to see where the butterflies "spent the night." She explained that the butterflies visited Saltillo every year on their way to the roosting sites in the mountains of Michoacán and México. "That's how we gave shape to the Correo Real program, and we started working on an activity manual so that teachers could touch on the theme in the classroom," she said.

Visiting students were enchanted by the story of the monarchs. Treviño would place a monarch on a visitor's hand and demonstrate how to tag and distinguish males from females. Immediately after their visit to the university, the children became guardians of the monarch, she said years later. For days after their visits, she would receive phone calls from the students, alerting her where monarchs had been spotted.

Treviño felt compelled to educate the community, but it was her friend and colleague Susana Mendoza, a French teacher, who came up with the idea of developing a program aimed specifically at teachers and students. The goal: instill understanding and appreciation of the monarchs and their migration.

To get started, the women perused the 1992 edition of the "green directory," a resource assembled by the office of the Secretary of the Environment of Mexico. The catalog included contact information for environmentally aware businesses, individuals, and teachers. They also reached out to the teachers who had attended the informal tours of the campus over the years. In a letter, Treviño and Mendoza invited prospective participants to join a quest to figure out the monarch butterflies' route through Mexico and help educate students in general about the monarch butterfly migration. Twenty-two people responded, and El Correo Real ("The Royal Mail"), one of Latin America's first citizen scientist programs, was born.

Soon volunteers were sending updates via phone, telegram, postal mail, and fax. Treviño collated the information into bimonthly bulletins to be mailed to her growing network of correspondents. The effort was driven 100 percent by volunteers. Hundreds of letters piled up on Treviño's dining room table every two weeks. The typed and handwritten salvos shared the date, time, place, weather condition, and count of butterflies observed for a particular sighting. Treviño recorded the data and noted locations on a map. She would then send summaries back to her correspondents, not unlike what Fred Urquhart had done annually in decades prior.

To help cover the costs of postage and printing, Treviño requested ten stamps from willing participants. People donated envelopes and paper. Photocopies were made at the PROFAUNA offices. Neighbors often pitched in, stuffing envelopes and licking stamps. Sometimes Treviño gave a *pellizco*, or "pinch," to the budget, borrowing from other projects or from her own pocketbook to pay expenses. In some cases, volunteers in Monterrey and

Coahuila offered to receive boxes of the letters and distribute them to schools on behalf of El Correo Real.

In the early years, the U.S. Fish and Wildlife Service supported Correo Real financially. Eventually, the state of Coahuila allowed the organization to use a government printer to produce the bulletins. By 2002, Correo Real was moving to the internet to distribute its updates. With the move to the web, participation fell from eight hundred correspondents to sixty. Some teachers and volunteers had no access to electricity, much less the internet. In some Mexican villages that is still true today.

In the early days, strong interest from U.S. teachers resulted in Correo Real's first teaching manual, *Mariposa Monarca, Manual Para el Educador Ambiental*, being translated into English. The translated booklet, *Monarch Butterfly, Manual for Environmental Educators*, served as an early example of monarch butterfly curriculum, and its content continues to be used today. It was one of the first teaching curriculums produced about monarchs in any language and was later adopted, utilized, and adapted by teachers in the states.

In the mid-1990s, the U.S. Fish and Wildlife Service invited Treviño to a meeting of monarch butterfly conservationists in Washington, D.C. There she met monarch butterfly scientists and advocates Lincoln Brower, Chip Taylor, and Elizabeth Howard, founder of the U.S.-based citizen science project Journey North. Treviño and Howard hit it off, and a long-term collaboration resulted.

The 1998 edition of the teaching manual, Correo Real's fifth, includes forty pages of monarch butterfly life cycle lessons, poetry exercises, crossword puzzles, illustrations of monarchs' body parts in all their stages, and a data collection sheet. The data form requests time of day, weather condition, an estimated number, and how long monarchs were observed—flying, nectaring, resting. The information gathered became the data set that eventually pieced together the insects' path through northern Mexico.

The teachers' guide also posed "Dilemas," critical thinking exercises, as part of the lesson plan. Scenarios posed in the manual forced students to contemplate the sometimes-impossible predicaments presented by efforts to conserve the monarch butterfly migration.

One example: Imagine you are a forest ranger at the monarch butterfly reserve. Your job is to prevent illegal logging. Given that your parents cut down trees at will many years ago to provide for an education that led to your current job, what should you do when you encounter old friends in the forest who intend to cut down trees?

What about the group of scientists studying the monarchs in the sanctuaries who think tourism is a viable source of income for local people and would prevent logging? But with more visitors, the community is cutting down more trees to build *cabañas* to house them during their stays. What to do?

And what if you are the owner of a logging company that operates where the monarchs roost? You provide work to many people in the area, but the prohibition of logging in some zones has forced you to fire workers and cut wages. You'd like to continue harvesting wood and creating jobs, but you know that adversely affects the monarch butterflies. What should you do?

Such were the ethical and practical quandaries posed as food for thought for young Mexican students in Correo Real's teachers' guide to monarch butterfly learning in the early 1990s. The "manual for environmental educators" assembled by Rocio Treviño more than a quarter-century ago remains relevant today.

To date, Rocio Treviño has received more than 15,000 mailed reports from Mexican citizens. As of this writing, she continues to oversee the citizen science project from a modest office in her home, mostly with volunteers and a small staff. In recent years the organization embraced What'sApp, a free, internet-based, global messaging service used widely in Mexico, to receive reports. To those who cross her path, Treviño is affectionately known as Señora Monarca or Abuelita Monarca—Mrs. Monarch or Grandmother Monarch. And Mexico's CONANP, the National Commission on Natural Protected Areas, lauds her as an "unsung hero" of Mexican monarch conservation.

Mexican media publisher Grupo Milenio, with funding from CONANP, construction giant OHL México, the Michoacán Tourism Institute, and Monsanto, released a documentary in 2018 called *Monarchs: The Spirit of the Forest, a Round-Trip*. The feature film, a first for Milenio, began in 2015 as a forum and series of videos and articles assembled by the multimedia company. Eventually the content morphed into an eighty-three-minute movie that chronicles the trajectory of the species through Mexico, a "legacy for humanity" of the monarch migration and the people of Mexico, according to Milenio.[6]

The project prominently features Treviño and credits her for her role in assisting with the initial mapping of the migration through northern Mexico to the high-altitude roosting sites. As the filmmakers make clear, the monarchs' path south of the Rio Grande was a revelation. The monarch migration story had traditionally focused on the U.S. leg of the journey or the monarchs' end game in the high-altitude Mexican forests. An important piece of the sojourn had been overlooked and left undocumented.

"When we hear monarch butterfly, we think of Michoacán or the state of México," said Roberto López, who codirected the film with Fanny Miranda and Miguel Galarza, in a promo of the series. But the butterfly does not teleport, he pointed out. It travels thousands of kilometers through Mexico—a trek largely revealed by the thousands of letters and sightings assembled by Treviño and Correo Real.

▼

Around the same time that Correo Real was mapping the migration south of the Rio Grande, another citizen science initiative driven by a woman was taking shape in the United States. Elizabeth Howard, founder and director of Journey North, had seen the stunning *National Geographic* story with Catalina on its cover when the magazine hit newsstands in 1976. Like many of us, Howard was moved by the color photo of the smiling Mexicana basking in a flurry of monarch butterflies. She read the fourteen-page article while a biology student at the University of Vermont. What she read in the captivating story amplified an awareness of the monarch butterflies that moved through Wayzata, Minnesota, where she grew up.

Howard and her sister, encouraged by their father, recovered monarch eggs from local milkweeds, brought them into the house, and raised the caterpillars in Skippy peanut butter jars. The impact of that *National Geographic* story on Howard was profound. She recalled that reading the evocative cover story of the "Monarchs Mexican Haven" was akin to learning something amazing about a friend you thought you knew well—but didn't.

In one of the most memorable anecdotes in the article, Urquhart had described a pine tree branch crashing to the ground "under its burden of languid butterflies." Unbelievably, he looked down and "by incredible chance" spotted a butterfly with a tag. That butterfly was tagged by Jim Gilbert, a longtime pal of Howard's, in Chaska, Minnesota, just thirty minutes south of Minneapolis.

More than thirty years later, Journey North, the organization born from Howard's passion and purpose, boasts more than 60,000 naturalists a year participating in tracking the monarch butterfly migration. Volunteers would heed Howard's call to "Go outside. Explore your own backyard. Get ready to share what you see!"

Journey North tracks wildlife migrations around the world—monarchs, hummingbirds, whales, eagles, and others. The monarch butterfly tracking initiative is the organization's most popular. Through it, Journey North promotes authentic experience as the greatest teacher by collating sightings of eggs, caterpillars, and adults through its website and the Journey North app. Participants' observations are posted in real time on an easily accessible, real-time, ever-changing online map, allowing people to see their contributions as they are made. Journey North also develops teacher resources and produces several weekly newsletters.

For decades all this unfolded from Howard's hillside home in Vermont, which overlooks Lake Champlain and the Adirondacks. Her path to Journey North had been both predictable and unexpected. A born nature lover, she spent years doing field research after graduating from the University of Vermont in 1975 with a biology degree. She worked on a moose study, researched Caspian terns, and monitored tigers in Africa. She also worked at the Nature Conservancy as a field steward and helped with fundraising. The field experience gave her credibility that can bridge the gap between academics and recreational naturalists. It has also informed her belief in experience-based learning. When her children arrived in the early 1980s, the family raised monarch butterflies at home, combining personal experience with metamorphosis and nature's magic.

In the early 1990s, the internet was just starting to move into the mainstream. Howard was early to the party for someone unschooled in computers. She became intrigued by the intersection of education and technology when a friend, Jennifer Kimball Gasperini, was working on an internet project that tracked the Arctic expedition of Will Steger. The adventure, chronicled on morning news programs and in newspapers, followed a six-person team, including two women, in the first dog-sled trek across the North Pole. The historic journey also engaged students via the internet in periodic updates shared with classrooms across the country in a program called Arctic Whiz Kids organized by Hamline University in St. Paul, Minnesota.

Howard had never heard of the internet, but upon learning of this novel education approach she mused that the idea could apply to wildlife migrations. And why not? Migrations include all the drama of survival stories—wrestling with the elements, searching for food and water, finding shelter. Wildlife migrations also touch the vast expanse of individual creatures that cross borders, cultures, and ecosystems. Such an initiative could be tapped to engage students in inquiry-based learning, which is knowledge acquired by the posing of questions rather than the memorization of facts.

Howard toyed with the idea and approached Gasperini to see if they could work together, marrying the Arctic expedition with wildlife. The idea held promise, and although the collaboration served as an incubator stage for Journey North the women could not secure funding. For two years Howard operated with no budget.

In 1994, Howard and her then-husband, Harry Roberts, flew from Minneapolis to Mexico City, rented a car, and found their way to Angangueo, Michoacán. The goal of the expedition was to establish a team and implement a protocol for sending live updates on the monarch migration to classrooms across the states.

Her destination was El Rosario, the largest, most popular, and most accessible monarch butterfly roosting sanctuary. The Monarch Butterfly Biosphere Preserve had been established just eight years earlier, in 1986. By the time Howard arrived, El Rosario was gaining credence as a relatively accessible ecotourism destination, one of the few places where tens of thousands of visitors could witness the magnificent roosting sites of the monarchs they had read about in *National Geographic*. In 1994, Howard and her husband were two of them.

She had no idea of the logistical and technical challenges they would encounter upon arrival in the remote mountain village perched on a steep hillside. Neither she nor her husband spoke Spanish. Neither had ever been to Mexico. Still, Howard jumped in. She recalled her brother-in-law chiding her with the taunt, "Ready, fire, aim"—her practice of diving in first, figuring out how to swim later. One thing she did know, however, was that if they could establish a team and a protocol for sending live updates on the monarch butterfly migration to classrooms across the United States, it could serve as a proof of concept.

The adventure connected Howard with Estela and Rosa Romero, owners of the Tienda Romero in downtown Angangueo, in the heart of the Monarch Butterfly Biosphere Preserve. The Romeros' corner storefront had hosted and assisted monarch researchers Brower and Calvert in the 1970s when the pair defied the Urquharts and shared the specific location of the monarch butterfly roosting sites with the world. Brower had made the introduction. As of this writing, Estela Romero continues to serve as Journey North correspondent in Michoacán. She shares photos and news of the monarchs' spring departure and fall arrival in the organization's weekly bulletins, which are shared and translated into Spanish for distribution by Correo Real. Throughout the 1990s, cowboy entomologist William Calvert was also a frequent contributor.

Howard spent a week in Angangueo coaching the Romeros on the basics of DOS computer language on a hefty desktop Hewlett Packard she and her husband had schlepped on the plane from Vermont. She also showed the Romeros how to hook up a modem. The idea was to have a team on the ground in Michoacán that could capture students' attention, just as Will Steger's Arctic expedition had engaged followers on his dog sledding adventure. Through a partnership with the Center for Global Environmental Education, an organization founded by Gasperini at Hamline University, the Romeros' updates were shared with four hundred classrooms across the United States.

On March 15, 1994, the first Journey North monarch butterfly bulletin was disseminated by fax. Alfonso Alonso-Mejia, a doctoral student at the University of Florida who worked with Lincoln Brower, sent the report, titled "Mariposa Monarca #1." In it, he introduced himself and explained that the research team was evaluating the microclimate of migrating monarchs. He discussed general monarch biology. The researchers were especially curious about the butterflies' mortality: do those that roost closest to forest clearings suffer more bird predation than those clustered in dense forest cover? The scientists were also looking closely at the structure and permanence of the forest, trying to determine their effect on the butterflies.

Alonso-Mejia provided a few more details before closing with the promise to return in a few weeks with reports of the butterflies' departure north to start the migration cycle anew. When he followed up later in March, others followed suit, reporting their FOS ("first of season") sightings of monarchs in the United States to Journey North.

About a year later, Globe TV, a Mexico City–based TV production company, documented the first live report of the monarchs' departure from Mexico. Mike Cerre, a friend of Howard's, accompanied Fernando Luis Romero, a fifth grader at Angangueo Primary School and the son of Estela, to the roosting sites. The story, "Monarch Migration from Mexico Begins!," included footage of their butterfly encounters for Globe TV, which was shared via the internet from Journey North's official HP computer–powered base station at the Romeros' store.

As often happens when visitors first experience the roosting sites, Cerre mistook millions of immobile monarchs for tree bark. When resting with their goldish-brown wings folded, the butterflies blend imperceptibly into the oyamel trees, appearing as dead leaves. "At first we thought the trees were dead, because they were rust orange instead of green," wrote Cerre. "As the sun warmed the forest, the monarchs came alive and started to fly. Within the next hour, the sky was filled with a blizzard of orange."

Cerre's bulletin and footage of the roosting sites aired on *Good Morning America*, a popular morning news program broadcast on the ABC television network. For more than twenty years after that TV exposure, Journey North had no need to hustle for publicity or funding, said Howard. The Annenberg Learner Foundation adopted the program.

Educators loved it. In the mid-1990s, content was assuming its throne as king, and back then there was not enough. Schools across the country were asking, How can we make use of this powerful new tool to enhance learning? Student contributions of data, information, and observations populated early web pages with much needed content that could be easily accessed by anyone with a computer and a dial-up connection. The observation, collection, and reporting of that data served a double duty as a learning exercise. It was brilliant. With an initial challenge grant from the National Wildlife Foundation and full funding thereafter from Annenberg Learner, Journey North grew into the premiere citizen science project that it is today.

In 2018, the organization's funding entered its sunset phase. For several weeks Howard feared Journey North might go extinct. But then monarch butterfly scientist and champion Karen Oberhauser, cofounder of Monarch

Joint Venture, stepped in and assumed its mission. Today, Journey North resides at the University of Wisconsin.

▼

Monarch Joint Venture, considered the United Way of monarch butterfly conservation organizations, also found its footing via the curiosity and tenacity of a woman scientist who was struck by the monarch butterfly's capacity to engage young people. In this case, the birth of a broad-based citizen science initiative resulted from an innocent question Karen Oberhauser posed to her daughter Amy's kindergarten teacher in 1989. "Would you like me to bring some caterpillars to your classroom?" Like any good educator looking for fun and interesting ways to bring science to students, the teacher responded, Yes, please—and can I have some for my colleagues, too?

At the time, Oberhauser had hundreds of monarch caterpillars and adults available from research she was conducting at the University of Minnesota. The Harvard-educated conservation biologist was investigating the biological investments made by males and females in reproductive processes. Specifically, she was researching how spermatophores, nutrition filled, gel-like sperm sacks, impacted successful reproduction.

Oberhauser had always assumed she would use birds in her study. But after learning how relatively easy monarch butterflies are to raise in captivity from an advanced field ecology course led by Chip Taylor of the University of Kansas, the pragmatic scientist changed her species focus to monarch butterflies. Soon she was collecting eggs, caterpillars, and adults in the wild for a controlled rearing operation in her lab at the University of Minnesota.

Her surplus livestock often ended up in classrooms at her daughter's school. The wonder and interest inspired by the chubby striped caterpillars and the glistening green chrysalises spawned the idea that monarch butterflies could be used to teach science to kids. What better way to educate young people about metamorphosis and basic ecosystem concepts than through the firsthand experience of raising monarch butterflies from egg to adult on milkweed?

By 1992, Oberhauser was working with local teachers and Elizabeth Goehring at the University of Minnesota to assemble a monarch butterfly–oriented curriculum for kindergarten through twelfth grade. Monarchs in the Classroom was born.[7]

The program became a flagship, the first of dozens to result from Oberhauser's deepening interest in monarchs. Her laser focus, penchant for collaboration, and ability to build successful partnerships resulted in myriad initiatives, including milkweed monitoring of monarchs in all their stages via the Monarch Larvae Monitoring Project; a sidebar nonprofit called the Monarch Butterfly Fund; an organization dedicated to assisting conservation and education efforts in Mexico called the Monarch Butterfly Sanctuary Foundation; and Monarch Lab, a University of Minnesota–based initiative that provides information, teacher training, and curriculum on the ecology, behavior, and evolution of *Danaus plexippus*. In 2008, Oberhauser forged an official partnership between the U.S. Fish and Wildlife Service and the University of Minnesota to form Monarch Joint Venture, a coalition of federal and state agencies, nongovernmental agencies, and academic institutions that work to protect monarchs and their migration through research, conservation, and education.

Monarch Joint Venture would likely not exist without Oberhauser. Her fingerprints are on every aspect of the collaboration, from its genesis in 2008 with ten organizations to its status as a legitimizer of myriad monarch butterfly–oriented nonprofits and NGOs.[8] An affiliation with Monarch Joint Venture gives an organization instant credibility. It's difficult to talk to anyone taken seriously in monarch science without Monarch Joint Venture's name being bandied about.

In addition to her nonstop advocacy of the monarch butterfly, Oberhauser has continued to research and publish hundreds of papers on *D. plexippus*. Her combined contribution to our understanding of the monarch and its migration has earned her deserved recognition, including a White House Champion of Change for Citizen Science Award in 2013. The national award recognizes her critical research on the habitat and conservation of monarch butterflies and the tapping of citizen scientists to collect data over long periods and across broad geographic areas.

With more than one hundred organizations under its umbrella as of this writing and its slew of scientists, administrators, and nonprofit advocates on its board and steering committee, Monarch Joint Venture assigns legitimacy—and sometimes small grants—to its member organizations. Funds are awarded through collaborative committees overseen by a steering committee of academics, citizen scientists, and representatives from government agencies and nonprofit organizations.

Monarch Joint Venture has succeeded, largely because of its ability to morph to the next stage. This includes ceasing the very practice that made it so captivating at its inception—making live monarch caterpillars and butterflies available to teachers and the public. Oberhauser was seduced by monarchs through her close contact with their magical life cycle. This has happened to many of us. But she grew to consider mass rearing irresponsible. She now recommends that teachers and students collect monarchs from the wild rather than purchase or import them from commercial breeders or university labs. She announced the policy change with a terse statement on her website in late 2007: "No Live Monarch Distribution in 2008."

"I've always had this sort of tension in me thinking, you know, it's not a completely positive thing that I'm doing," Oberhauser told reporter Adam Federman of *Earth Island Journal* at the time. "I always felt that tension."[9]

High mortality rates associated with moving her lab to a new location on campus gave Oberhauser pause and led to the change. She had serious concerns about unleashing 50,000 monarchs into the wild each year, possibly spreading disease from her lab to the migratory population and the roosting sites in Mexico. Though she recognized the tactile experience of raising and releasing monarchs as a unique opportunity to spur the public's imagination and even embrace action, Oberhauser concluded that from a conservation perspective mass rearing was inappropriate. Besides, capturing and raising local monarchs from the wild would result in more authentic learning—*if* you could find them.

This attitude shift evolved with monarch butterfly science. One of Oberhauser's own protégés, Sonia Altizer, was largely responsible for spearheading the science that raised the worrisome red flags. A distinguished professor at the University of Georgia's Odum School of Ecology, Altizer was one of the first scientists to study *Ophryocystis elektroscirrha*, or OE, as it's known in monarch circles. The spore-driven, protozoan disease infects monarchs and other butterflies that use milkweed as their host plant. OE is present naturally in monarchs, just as the *Streptococcus* bacteria is prevalent in human beings. And just like strep, OE can cause problems in an accommodating environment or weak host organism. When OE gets out of control, it can cause crippling or death.

Monarchs must consume the spores to become infected, and the spores are easily carried and transferred since they fall off butterflies like salt from a shaker. Altizer's research suggested that monarchs raised in crowded environments are more likely to be infected. With the science in progress, Oberhauser decided that unleashing monarchs raised in a lab or commercial breeding operation that *might* be carrying OE was just too dangerous. It could sully the wild population and possibly place the migration at risk.

Altizer was a first-year grad student looking to study insect pathogens in Oberhauser's lab when Oberhauser was raising and distributing thousands of monarch butterflies in their various stages for her Monarchs in the Classroom program. That's when the rearing operation hit its disturbing snag. Monarchs were emerging dark and deformed from their gold-flecked, jade green chrysalises—black, mottled, obviously weak and with crinkled, deformed wings. Several hundred butterflies died. Baffled, Oberhauser summarily placed a mound of gooey, black monarch butterfly corpses on Altizer's desk. "Sonia, you like diseases. Can you figure out what's killing these butterflies?" she asked.[10]

The request set in motion years of OE research and ultimately led to considerable contention in the monarch world. The prevalence of OE challenged not only the appropriateness of mass butterfly releases but also the planting of a particular milkweed, *Asclepias curassavica*, tropical milkweed.

According to Altizer's research, the technically nonnative tropical milkweed, an easy-to-grow, orange-blooming perennial that is often the only *Asclepias* species commercially available, encourages monarchs to break their diapause and lay eggs, even in the fall when they are supposed to be migrating. (Read more about tropical milkweed and mass rearing in chapter 12.)

Oberhauser's sober passion is hard to miss when she looks back on her decades-long relationship with the monarch butterfly. She first ventured to the roosting sites in 1992 before the instigation of annual population counts. By 1996 she was leading her first crew of graduate students on a tour to Mexico.[11] That was the third year of the World Wildlife Federation's annual monarch butterfly census, when the population spiked to forty-five acres of occupied forest and was estimated at more than 900 million butterflies, the highest in recorded history. The 2007–8 population, when Oberhauser canceled her mass rearing program, represented a 75 percent drop from that peak—only about eleven acres, or 230 million butterflies, were counted.

"I've seen that whole decline," said Oberhauser. "I've seen it up here in the summer and down there in the winter." Oberhauser visited the roosting sites in their record year—1997–98. As a child, she had frequently seen monarch butterflies in Minnesota during summers as a kid. Her father, a veterinarian, and mother, a high school counselor, encouraged rearing them at the kitchen table, an activity she replicated for her daughters. "Everybody did," she said.

But that's more challenging these days with the monarch population still in decline. When Oberhauser's daughters attended school in the 1990s, monarch butterflies were regular Minnesota visitors during late summer and early autumn. In recent years, few monarchs can be found at back-to-school time.

In 2017, Oberhauser made a career change, a move almost as surprising as her decision a decade earlier to stop rearing monarch butterflies for teachers' use in the classroom. She resigned her post at the University of Minnesota to become the director of the University of Wisconsin Arboretum in Madison. Aldo Leopold, the father of wildlife conservation as we know it, once occupied the position she was moving into.

When asked about her decision to leave Monarch Joint Venture—the organization she helped found, has defined her professional life, and has made her name synonymous with monarch butterfly conservation around the world—Oberhauser cited a desire to be closer to family. Her daughter and grandchildren were living in Wisconsin at the time.[12]

Adapting her focus to prairie restoration also made pragmatic sense. It speaks to the belief that her thirty-year relationship with *D. plexippus* is not just about orange-and-black butterflies. It's about the entire ecosystem. Her 2019 adoption of Journey North, which tracks not only monarchs but hummingbirds, whales, cranes, and other species, reinforces the notion that, although monarchs are an effective ambassador, it's not just about the butterflies; it's about the whole of the planet.

Oberhauser believes devoutly in citizen science as a tool for creating a greater understanding of our world. She wants the natural sciences to be relevant again. Conservation biology is the "science of hope," she said. "Am I hopeful that at some point we're going to turn this juggernaut, the human domination of the planet, around?" she asked. "I think we have to. . . . I think we will. . . . Whether it will be in time to save monarchs, I don't know. But I wouldn't be in this field if I didn't have hope.

▼ ▼ ▼

CHAPTER 6

THE MONARCH WATCH
AND D-PLEX

Monarch butterfly tagging pioneers Fred and Norah Urquhart closed their research associate–driven initiative in 1994, just two years after Chip "Orly" Taylor, a fifty-six-year-old entomologist, took up the cause and launched Monarch Watch. The citizen science tagging program operates in the prime monarch butterfly summer breeding grounds of the American Midwest.[1] Its success would cast Taylor, whose signature scruffy white beard and blazing blue eyes call Santa Claus to mind, as the unofficial *padrino* of monarch butterfly citizen science.

The effort began as an experimental research project at the University of Kansas at Lawrence. In 1992, Taylor tapped Brad Williamson, a teacher at Olathe East High School. Williamson convinced Taylor that students and teachers would be interested in the tagging program and could provide high-quality data.

During the first year, Williamson had access to a printer and printed the tags. He and Taylor would meet at a McDonald's about halfway between Olathe and Lawrence. There, Williamson would hand over the tags, which Taylor then mailed from the university offices in Lawrence.

Together, Taylor and Williamson engaged five hundred adults and four thousand children to tag an estimated five to six thousand butterflies that first season. The pilot program utilized Urquhart's imperfect alar tagging system of applying tags to the leading edge of the butterflies' forewing with glue. Given its uncertain future, Taylor and Williamson's program didn't appear to have a name. The eight-page season summary Taylor drafted and mailed to volunteers in April 1993 was printed on University of Kansas Entomology Department letterhead. It channeled Fred and Nora Urquhart's familial style, reading like a personal letter and greeting its readers, "Hello Monarch taggers!"

"WOW! What a year!" Taylor started. "When we sent out news releases requesting help in tagging monarchs, we had no idea how much of a response we would get. We were overwhelmed!"[2] The public's warm embrace of monarch butterflies contrasted sharply with the general sentiment toward bees, the insect Taylor had been devoting himself to in recent decades. Specifically, Africanized "killer bees."

Working toward his PhD in zoology at the University of Connecticut in the 1960s, Taylor studied insect ecology and population biology. He realized as he wrapped up his dissertation in 1969 that he was severely allergic to butterflies of the Pieridae family, commonly known as sulphurs, which were the focus of his research at the time. It's not uncommon for scientists to develop severe allergies to the subjects of their study.[3]

By 1972, Taylor had become asthmatic and was using steroids daily. Then he experienced a dramatic episode of anaphylaxis, an extreme, scary asthma attack. His air passages constricted and his heart raced to the point that he thought he was going to die. When his condition escalated to the point of requiring regular doses of prednisone, he knew he had to make a change.

Two years later, the pragmatic Taylor switched his scientific focus to Africanized honey bees. The invasive insects were brought to Brazil in the 1950s in a move to boost honey production. More than two dozen swarms escaped quarantine at a São Paolo research facility in 1957 and began making their way north, cross-breeding with more docile native bees along the way. Entomologists throughout the Americas were keeping an eye on the hybrid bee, whose hyperdefensiveness and tendency to swarm earned it the foreboding and unfair "killer bee" moniker. Media reports played up the creatures' aggressiveness rather than their capacity for stellar honey production and hive growth. Africanized bees tend to produce 25 to 100 percent more honey than domestic strains, which is why they were imported.

With allergies ending his sulphur butterfly studies, Taylor smelled an opportunity. He turned his attention to the unwelcome interlopers. For two decades he churned out research on the Africanized honey bees, establishing research sites with students throughout the hemisphere. In the early 1990s, as scientific study increasingly tapped new technologies, he realized that to continue his

Africanized honey bee research he would be forced to wade into molecular biology, a field that didn't interest him. A gregarious man who obviously enjoys engaging with people and working in the field, Chip Taylor is hard to imagine confined to a laboratory, looking through a microscope. The prospect pushed him toward monarch butterflies. "I morphed twice," said Taylor.[4]

Taylor recalled a general awareness of the news that Urquhart's Insect Migration Studies Association was shutting down. This realization occurred just as funding for Taylor's honey bee research was coming to an end. He took a look at the Urquharts' research and surmised that the only thing Urquhart appeared to be truly interested in was *where* the butterflies went. That mystery had been solved in 1975. And yet . . . so much could be learned from tagging butterflies, recording their data, and monitoring their flight patterns, habits, and tag recoveries in Mexico. Taylor decided to turn his attention to the science associated with migrating monarch butterflies.

The experiment that started at Olathe High School has endured for more than three decades and evolved to become the premiere engagement tool and one of the most influential institutions in monarch butterfly citizen science. Taylor's independence and pragmatism explain much of the institution's success.

In addition, the powerful draw of Monarch Watch comes from the utter novelty of its signature tagging program. The idea of tagging a butterfly never ceases to spawn a series of questions, beginning with "how do you do that?" Monarch Joint Venture has its monarch caterpillar monitoring project, and Journey North and Correo Real encourage the observation and sharing of monarch sightings throughout the season through the web, mail, and phone apps. But the tactile contact with the species associated with the Monarch Watch tagging program holds irresistible and unique appeal. The tagging and release of *Danaus plexippus*, combined with the magic possibility of an individual butterfly's recovery thousands of miles south, captures the imagination. With tagging as its centerpiece, Monarch Watch has lured tens of thousands of novice naturalists to engage in pollinator conservation.

It took some time for Monarch Watch to find its footing. By May 1995, the newsletter representing the "outreach program that we now call the Monarch Watch" arrived with a large "MONARCH WATCH" header. A chubby, cartoonish, striped monarch butterfly caterpillar, a pair of binoculars perched clumsily

Chip Taylor visits the Texas Butterfly Ranch on the Llano River in October 2012.
Photo by Monika Maeckle.

atop its antennae, looks skyward. As of this writing, this primary logo still graces some pages of the Monarch Watch website, despite a rebranding in 2013. In the tradition of Fred and Nora Urquhart, Monarch Watch outsourced monarch butterfly research, bringing citizen science and the magic of monarchs to tens of thousands of students, novice lepidopterists, and nature lovers from Mexico to Canada.

In the early days, Taylor struggled with the tags, just like his predecessor. Legible, enduring, practical tags that would cling to a butterfly through dry, wet, hot, and cold over many months eluded him for several years. From 1992 until 1998 the program used the adhesive-applied alar tag method pioneered by the Urquharts. In the April 1994 newsletter, Taylor mentioned that tagging kits would include twenty tags, instructions, and a "vial of adhesive for tag application." Tags were sent free of charge to "collaborators," the word Taylor used in place of Urquhart's "research associates."

Back then, Monarch Watch tags still had to be adhered with glue—an imperfect process. And no matter how explicit the instructions, collaborators always used too much glue, inadvertently getting butterflies stuck to their fingers or the lawn furniture, gluing wings together, and sometimes seriously damaging the butterflies in the process. Jenny Singleton, who first engaged in tagging in Grapevine, Texas, when the adhesive system was still in place, characterized the antiquated alar tagging system as "a mess." And recovery rates were miniscule because the tagged butterflies were often damaged or the tags fell off.

Taylor knew the tagging system had to change. But how? He and his team continued to experiment. They tried rectangular tags for a while—but the corners kept curling up and falling off. A circular tag would be better, he imagined. No corners. Eventually he contacted Watson Label Co. in St. Louis, Missouri, to see about developing a special tag that was easy to adhere and would endure the insects' arduous, multiseason journey.

Then there was the adhesive. Taylor spoke with Wayne Roth, a senior sales rep in the industrial tape and specialties division at Minnesota Mining and Manufacturing Co. in St. Louis, the company that would later become 3M Corporation. Roth, intrigued by the special requirements of the tag, worked with adhesive engineers to satisfy the tag's unique needs.

The glue had to be thick enough to seep through a layer of scales and stick to the wings. It couldn't be toxic to the butterflies. It required a special die cut and would need to be affixed to another background sheet from which it peeled off easily. Each tag also required a unique code, number, and University of Kansas imprint. What appears to us today as a nondescript, common sticker was quite specialized for the mid-1990s, necessitating a singular tool, a unique adhesive, and a special backing. "Not a trivial technology in any way," Taylor said later.

The 3M team collaborated with Watson Label Co. and utilized adhesive 9442 because of its durability.[5] After several trials, Monarch Watch settled on the tag used today—a circular, white polypropylene dot printed on all-weather stock with permanent ink. The tags, printed on four- by five-inch sheets and sent by mail with detailed instructions each August, measure 0.89 centimeters in diameter and weigh less than 0.01 grams.

More than thirty years later, Monarch Watch sells about 200,000 tags annually to volunteers across the country. Ordering tags is a late summer rite. Many of us assess the season, wonder if it will be a good year or a bad one,

Tagged monarch on the Llano River. Photo by Monika Maeckle.

and budget how much time we'll be able to devote to tagging outings before placing our orders. Tags cost $15 for twenty-five tags, $50 for two hundred.

Monarch Watch staff estimate that only about half or fewer of the tags sold each year are utilized to tag monarchs. Typically, an average of 1,500 (1.5 percent) tagged butterflies are recovered at the sanctuaries in Mexico. Since 1992, volunteers have tagged more than 1.4 million monarchs, and more than 17,000 of the tagged butterflies have been recovered. I've lost count of my own personal tally, which I occasionally reviewed at the Monarch Watch Recovery Database before it went offline in 2015. More than fifty-five of around eight thousand I've tagged have been recovered and reported. Singleton, who serves as a docent at the Grapevine Flutterby Festival and at her family ranch in Hext, Texas, has tagged more than 15,000 monarchs over the years. She has seen at least one hundred recoveries.

Until 2015 the recovery database was easily searchable on the Monarch Watch website but has since been taken down to "clean up the data," according to Taylor. Now only the most recent season's recoveries are posted. Not as accessible, but a temporary situation, he said. Social media and email have moved to fill in the gap of identifying recovered tags. Various Facebook pages and email listservs are devoted to sharing photos and info about tagged monarchs recovered in the United States and elsewhere.

Monarch Watch representatives say they hope eventually to digitize the data at the time of tagging so that notifications of recoveries can be electronically automated in real time. The organization started making digital spreadsheets available for entering tagging data in 2012. But much of the data continues to be returned on paper and must be transcribed.

In the 1997–98 season summary, the first year the new style of adhesive label tags had been employed on an experimental basis, Taylor touted Monarch Watch's best tagging year up to that time. The organization had issued more than 192,000 tags and estimated that 76,000 butterflies were tagged. More butterflies were recovered that year than in Monarch Watch's two best years of tag recoveries combined. The 1997–98 season was also the first year that Monarch Watch paid for tag recoveries. Local people at the roosting sites were paid $5 per recovered tag, an incentive that surely increased recoveries.

Other developments boosted Monarch Watch's credibility and profile in the 1990s. In 1993, Taylor recruited William Calvert, witness to the "unpleasant incident in the forest," to the Monarch Watch citizen science effort. The iconoclastic Calvert provided interesting context from Texas, where two major monarch flyways were identified for the first time. One flight path ran through the Texas Hill Country in the middle of the state. The other hugged the coastline along the Gulf of Mexico. Together these two pathways constitute the renowned Texas Funnel, through which all migrating monarch butterflies must pass, moving to and from their winter roosts.

Taylor, like Urquhart, understood the power of the media. From the earliest days, both men used their newsletters to tout the media placements, mostly in newspapers, that carried their messages of monarch butterfly engagement to the masses. In 1994, Taylor listed the regional newspapers in which "the project"—unnamed at that time—appeared: the *Kansas City Star*, *El Dorado Times*, *St. Paul Pioneer Press*, and *Omaha Herald*. His keen understanding of how the press can serve as ambassador to the cause has served him well through the decades.

But perhaps most consequential was Monarch Watch's early adoption of a web page and listserv technology—a "Monarch Archive on the Internet" and a "Monarch List" through which participants could share information and pose questions. The latter, first mentioned in 1995, became the glue that continues to bind the far-flung monarch butterfly tagging community.

"Monarchs Take the Information Highway!" Taylor announced in his 1994 seasonal summary. Today, the organization's explanation of its embrace of the internet reads like a page from *Internet for Dummies*. Here's an excerpt: "This fall, we began an archive of monarch biology and information about Monarch Watch on World Wide Web. WWW is an information service on the Internet that supports text, pictures, sound and movies. Our archive is called a 'homepage' and is a document to WWW that has text and pictures (no movies or sounds yet, but there will be!)."

The article proceeded to explain the possibilities of "'hypertext,' which presents information where selected words in the text are linked to other documents," as well as how one can access the web through a "'browser,' which is a program that allows you to view homepages served by remote computers all over the world!"

So, in addition to enlightening the masses on the whys and whims of the monarch butterfly migration, Monarch Watch also helped internet newbies carefully tread, perhaps for the first time, the World Wide Web. Of course, to do so one would need an electronic mail (email) account, the newsletter advised.

The article served as the birth announcement of the D-PLEX-L, or simply the D-PLEX, as we who follow it call it. Short for *Danaus plexippus*, the D-PLEX has become the online public square where citizen scientists, professional scientists, hobbyists, and other monarch aficionados with varying degrees of seriousness learn from each other in real time via email. The community shares FOS sightings, exchanges tips on how to germinate milkweed seeds, and debates environmental policy. On any given day, an argument over the pros and cons of planting tropical milkweed or engaging in commercial butterfly releases comingle with pleas to share local ecotype milkweed seeds or announcements of monarchs' arrival to their northernmost migration point. Politics are not verboten on the D-PLEX, but at times testy exchanges have escalated to the point of requiring intervention by Taylor himself. Several people have been banned from the D-PLEX for inappropriate interactions, hijacking conversations, or outright rudeness.

The D-PLEX has a vigorous life of its own, and like an invasive species it can become seasonally hyperactive, clogging up your inbox. During migration season the list chatter bloats to dozens of exchanges a day; it tallies three to five thousand email messages per year—two and half decades after its debut.

Its vigor is remarkable in an age of email decline. Niche email lists often atrophy with time as people lose interest and social media groups become the preferred communication mechanism. Yet the D-PLEX continues. Conflict and differences of opinion fuel it. For years the annual migration update included a "best of the D-PLEX" feature, where provocative questions posed in email were recycled like best hits. "Sometimes Chip rises to these as trout to a fly," read the 1995 season summary. "Be patient with him. He gets excited about unanswered questions and tries to get you excited as well."[6]

The "best of" feature quickly moved from summarizing Taylor's musings and accounts of his most recent trip to Mexico or the latest vector studies to recapping some of the more esoteric knowledge from the crowd. Tips on how to move a monarch chrysalis safely and info on ridding your milkweeds of aphids educate list followers. This gem from Chuck Safris explains an ingenious way to create a low-cost butterfly net: "Anyone who is looking for some nice heavy wire to make a hoop for a butterfly net need look no further than the lawn signs of your favorite political candidate. Call your candidate's office and ask for a yard sign, then after the election, help yourself to about five feet of heavy metal wire . . . and don't forget to vote."

The D-PLEX continues to grow in the sophistication of its exchanges. PhDs, commercial butterfly breeders, and professional pollinator advocates mingle with the monarch masses and generally provide a high level of intelligent discussion. A spring exchange in 2018 debated the likelihood (or not) of monarchs achieving threatened status under the Endangered Species Act during the Trump administration. D-PLEXers jumped in with relevant news articles and forgotten-but-pertinent studies. All those who followed the thread walked away from their devices better informed.

One esoteric exchange had Taylor and next-gen monarch researcher Anurag Agrawal debating the veracity of a claim made in a 2006 article that female butterflies can be infused with so many spermatophores by their randy mates that their bodies burst. The serious but civil exchanges were extremely interesting to the nonscientist:

"I'd never heard of that," Taylor wrote politely. "One would think you would occasionally see the same thing in mating cages, but that's never occurred in our facility and I've never heard monarch breeders talk about that. Interesting."

Scientists continue to study the many puzzling questions about monarchs. But perhaps more important than the 1.4 million tagging records in the data trove resulting from decades of citizen science is the coinciding engagement. When tens of thousands of human beings connect intimately with one of the most charismatic species and remarkable natural events on the planet, something special occurs. Personally, I doubt I would have mined my own interest in monarch butterflies and their migration were it not for the tactile interspecies connection that accompanies netting, handling, tagging, and releasing the butterflies.

How long the tagging program continues in its current form remains to be seen. Scientists and technologists are collaborating to figure out a better way to tag monarchs. The manual application of a sticker to the wing of a butterfly seems adorably quaint—akin to licking stamps—in an era of drones and microchips. And the current method, for all its attributes, fails to provide key data points, such as the butterflies' specific migratory habits and how environmental conditions affect their flight. Scientists continue to seek answers, which will shed light not only on the monarch butterfly migration in particular but conservation, climate change, and sustainability issues in general.

In late 2017 the Monarch Butterfly Fund, a nonprofit organization devoted to monarch conservation in Mexico, announced an international design competition to develop better technology to track and relay data on monarch butterfly migration habits. The winning design could potentially earn a $50,000 grant. Although no single entry met all the qualifications, the Fund eventually awarded a team from the University of Michigan $10,000 in response to a proposal to develop tiny radio transmitters to be adhered to the backs of migrating monarch butterflies.

D. André Green II, an evolutionary biologist, partnered with his university's engineering department led by electrical engineering and computer science expert David Blaauw. The two developed tiny sensors that were used on migrating monarch butterflies in a pilot program at the Texas Butterfly

Ranch in October 2022. Research continues. Once the technology is perfected to allow the tracking of individual butterflies, "a lot of doors will be opened," according to Green, since the U.S. Department of Defense is already strapping tiny cameras to the backs of insects to act as mini-drones. In theory, this technology could be used in the same way.[7]

Other innovations include Project Monarch, a community science tracking app that uses bluetooth technology to follow monarchs outfitted with tiny transmitters the size of a grain of rice.

Monarch Watch, like its founder, will adapt. Taylor served on the board of the Monarch Butterfly Fund for years and supported the evolution of the tagging program. As was the case with Oberhauser, what began as an exclusive focus on *D. plexippus* has evolved into a more general conservation effort—an inevitable outcome of embracing the life cycle of the monarch butterfly. That's fine with Taylor, who sees the monarchs' plight as a warning for all pollinators and the ecosystems that sustain them—and us.

In the summer of 2022, Taylor shared his intention to retire at the age of eighty-five. He made the announcement at the thirtieth birthday celebration of Monarch Watch, an invitation-only gathering of professional and citizen scientists at Monarch Watch headquarters in Lawrence, Kansas.

After a day of remote and live presentations as well as a tagging outing, Taylor explained that he and his wife Toni would be committing $1.4 million in Apple Computer stock to the KU Endowment, the private foundation authorized to raise funds for the University of Kansas. The gift would be used to ensure the future of Monarch Watch by establishing the Monarch Watch Chip and Toni Taylor Professorship, a faculty position at the University of Kansas dedicated to overseeing the science of Monarch Watch.

In October 2023, Monarch Watch announced the appointment of its new director, Kristen Baum, a professor of integrative biology at Oklahoma State University. Baum's research focus has been on the effects of land use and management practices on pollinators, including native bees, honey bees, monarch butterflies, and other insects.

▼ ▼ ▼

CHAPTER 7

THE DROUGHT OF 2011 AND ITS BUTTERFLY EFFECT

When Catalina and Ken first walked the monarch forest in 1975, the butterflies likely numbered a billion or more. A little more than three and a half decades later, the entire eastern population of migratory monarch butterflies totaled an estimated 14 million.[1] In other words, all the migratory monarchs east of the Rocky Mountains could fit into the average Walmart store with 30,000 square feet to spare—a 90 percent decrease from the late 1990s when population counts were first conducted.

What happened? The historic Texas drought. Scientists often cite Texas as the most important state in the migration because it's often the first place where spring monarchs lay their initial round of eggs, and one of their last stops in the fall when they're fueling up to build up their fat stores for successful overwintering. Moisture in the form of rain is paramount to producing welcoming and healthy foliage in the form of milkweeds in the spring and sugar-laden nectar plants in the fall.

The lack of precipitation in 2011 set the stage for what some experts have deemed the driest year in Texas history. According to the National Weather Service, an average of 14.8 inches of rain fell on the Lone Star State that year, compared to the usual 32 inches. Public water utilities implemented strict restrictions on outdoor irrigation, which consumes more than 70 percent of the state's water each year. Cities across the state ruled to limit or completely shut down automatic sprinkler systems, used by many homeowners to keep nonnative turf lawns alive and green. Hose draggers, those of us who sate our yards by watering manually, could water any time, any day.

In the evenings, I looked forward to dragging my garden hose around my small front yard in Austin, Texas, where I lived for a year while opening a new office for my employer, the press release wire service Business Wire. Several evenings a week, I drenched my plants with a good, long drink. One, two, three, four—slowly I counted to ten, strategically soaking the primary stems of plants I had curated for transport and transplanting from San Antonio. Water seeped through several inches of mulch to the rich, black soil below. My new neighbors' St. Augustine grass lawns looked scorched and brown. But my butterfly garden rocked. Even during the historic Texas drought.

At the ranch, Monika's Pond, a newly constructed stock tank that had filled up with previous years' rains, evaporated. We had added several loads of bentonite to the tank bed to encourage a wildlife watering hole. The natural clay is the primary ingredient in kitty litter because of its ability to hold liquids. At the beginning of the drought, the tank became a muddy mosh pit and open invitation for feral hogs. Before all the moisture disappeared, the omnivorous pigs stubbed their snouts into the damp earth in search of grubs and roots. They left a disturbed, tilled tank bottom in their wake. By June, the "pond" had become a dry bed of grayish-white bentonite and soil, marked by conspicuous half-inch-wide cracks. Just months before, we had snapped a silly photo of me in my kayak, paddling in circles.

The Llano River, the lifeblood of our small ranch, dropped precipitously over the summer. By August we could walk on the karst riverbed to the Chigger Islands without getting our feet wet. A new limestone path jutted up from the anemic water flow. Weekend visits left us wondering, What if the river dries up?

One Sunday afternoon upon returning to Austin in September, I drove into town as fire engines doused the north side of Highway 29. The frontage prairie of a horse farm had been scorched by fire, leaving charred earth in its wake. The field was still smoking as I drove past. Not a single stalk of green remained.

When this dry stretch began in the fall of 2010, no one thought much about it—Texas weather, just wait. It'll change soon. But then the rains didn't come and didn't come. By spring we began to understand that it would be an exceptionally hot, dry summer like none before. In fact, some believe the 2011 drought left Texas more parched than the Dust Bowl of the 1930s.

Monarchs and other butterflies visited my garden throughout that spring, despite the lack of rain. The Lady Bird Johnson Wildflower Center predicted

First instar monarch caterpillar on antelope horns milkweed.
Photo courtesy Chuck Patterson

an unimpressive spring wildflower display for 2011. On a ranch outing in April, the bluebonnets, Indian blankets, and paintbrush that typically dot the roadsides and draw flower gazers from across the state that time of year were unusually sparse or absent. Governor Rick Perry declared the weekend of April 22, 2011, the official "Days of Prayer for Rain in the State of Texas."[2]

When migrating monarchs arrived from Mexico in March, the only wild-flowers visible on many Hill Country outings were antelope horns milkweed, *Asclepias asperula*. This drought-tolerant native managed to eke out a few visible but modest stems, its odd green and white blooms projecting from the thirsty soil. Their presence constituted a lucky coincidence for the butterflies since they could lay eggs *and* nectar in a single visit.

By September the heat had taken a devastating toll. The late-season nectar blooms that typically show along the Llano were scorched. Burned expanses of brown biomass replaced the usual joyous flowers.

Frostweed, a fascinating and underappreciated plant so named for its ten-dency to split its stem at the first frost and ooze its phloem to form amazing mini-ice sculptures, was among the drought's carnage. The reliable white flower is a Texas Funnel favorite for migrating monarchs each fall. It typically flaunts its showy blooms starting in September, just as the first monarchs arrive.

In 2011 the dependable nectar source was parched dead in its tracks. Four-foot-tall stems stood erect, brown, and crisp along the banks of the Llano River. One hot dry weekend in late September, a handful of monarchs mingled with dragonflies, occasional bees, and a lonely eastern swallowtail as the random skipper flitted past. A solitary white-blooming snow-on-the-prairie pushed its way from the cracked soil, just a few yards from the river bank, where goldenrod struggled to produce stunted blooms. Pink swamp milkweed umbels jutting from the Chigger Islands hosted loads of monarch eggs from the premigratory wave that moved through around Memorial Day. But the plants measured half their usual size as the river continued its sad recession. A grove of mature pecan trees, always a favorite monarch overnight roosting site, showed serious drought damage. Major limbs lay on the ground, felled from their own dead weight.

We were deep into these parched days when monarch butterfly expert Lincoln Brower acted on a hankering to jump on a plane to Texas and witness this historic drought and its impact on migrating monarchs firsthand. Brower flew into San Antonio the week of October 7, 2011, just as peak monarch butterfly migration season began in Texas.

At age eighty, the famous lepidopterist planned to grab samples of migrating monarchs to measure their fat deposits as they moved through the Texas Funnel. His theory: the butterflies arriving in the Lone Star State this year would be scrawny and compromised. Their requisite stored fats would be prematurely depleted by the drought, Brower proposed, making them ill equipped to make it all the way to their roosting sites in Mexico.

Overwintering monarch butterflies rely on stored fats to fuel their long journey and winter stay in the Mexican mountains. Just as hibernating bears eat extra honey, nuts, and berries to put on weight that will carry them through the winter, migrating monarchs consume carbohydrate-rich nectar from late-season flowers. By the time they arrive at the high-elevation forests, they've been loading up for weeks. Those excess carbs convert to lipids, or fats, which they store in their abdomens. And those fats make it possible for the butterflies to survive their months-long winter stay in Mexico.

Compounding the nectar shortage in the Texas Funnel, more than 31,000 wildfires had raged throughout the state in 2011, burning 4 million acres and destroying almost three thousand homes. Their smoky haze and burning heat added another challenge to migrating monarchs, while some Texas weather stations reported less than a tenth of an inch of rain for the entire year.[3] According to the Lower Colorado River Authority, the twelve months from October 2010 to October 2011 marked the driest spell on record since record keeping began in 1895. And the Texas drought map, assembled weekly by the National Weather Service, showed more than 70 percent of the state in "D4," burned, red shorthand for exceptional drought. Later, hydrologists would determine that the so-called drought of record in the 1950s was still the worst in history, but 2011 ranked a close second.

As Brower's visit approached, Kip Kiphart, a longtime volunteer and trainer for Monarch Joint Venture in the San Antonio satellite community of Boerne, called to see if I might be interested in joining an outing to collect butterflies for Brower's study. The team needed access to private ranches in the monarch flyway—could I help?

Of course! I called my monarch mentor, Jenny Singleton. Having tagged in the area for more than a decade, she knew all the butterflies' favorite fall haunts in Mason, Kimble, and Menard Counties. Over the years, Jenny had cultivated several private landowners, a potentially testy group of folks who generally don't take kindly to strangers with butterfly nets trespassing on their land without permission. Jenny knew locations traditionally associated with monarch butterfly roosting and nectaring. During the migration season, she often spent weeks at a time on her ranch, interrupting her stay with regular explorations of the Hill Country. Jenny knew that monarchs, like other wildlife, naturally gravitate to water. The area's streams, rivers, ponds, and creeks were the likeliest, often only, places that could serve as nectar pit stops this time of year, especially during a drought.

Jenny and I joined Lincoln Brower's Texas Hill Country Drought Tour as the citizen science contingent. We both were honored to spend time with the legendary lepidopterist. The butterfly-chasing dream team also included Mike Quinn, an Austin-based entomologist and founder of Texas Monarch Watch and the Austin Butterfly Forum, and Kiphart, a retired cardiologist

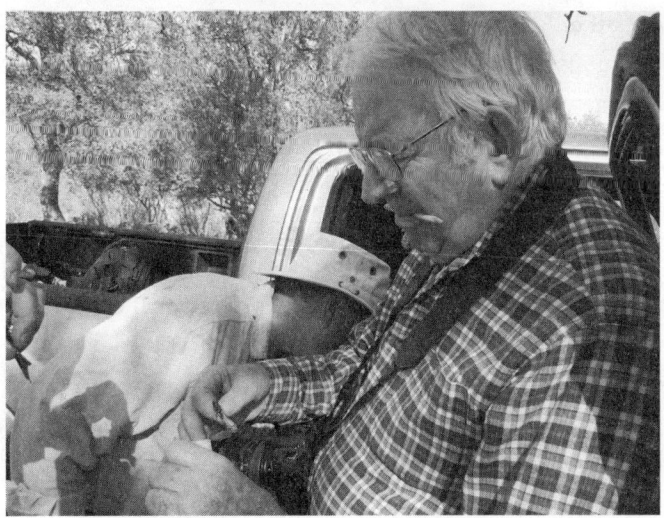

Lincoln Brower collects monarchs for his lipid study on the Llano
River in 2011. Photo by Monika Maeckle.

and award-winning manager/trainer for the Monarch Larva Monitoring
Project, an offshoot organization of Monarch Joint Venture.

We started the day at our place on the Llano River, between Mason and
Junction. A cloudy sky shrouded the karst watershed as we hiked from the
house. Not much was blooming. Across the river, hundreds of dead Ashe juniper trees dotted the bluff like burnt orange beacons, victimized by drought.
I explained to Brower that in Texas we call them "cedars," even though the
ubiquitous Ashe junipers technically belong in the cypress family.

Few monarchs were flying on our stretch of river, but we netted six. We
handed our catch to Brower, who appraised each one. "Skinny." "Fat." "She
looks pretty good." "Porker." He took copious notes in an old-school notebook while deftly folding the butterflies into glassine envelopes, protective
waxed paper enclosures, before storing them in an ice chest. Later, at his
laboratory in West Virginia, he would grind them into a paste and measure
their body fat percentages.

Brower shared a handy trick for those who tag butterflies: how to determine
males from females without unfolding their wings. Checking a butterfly's
sex can be an awkward maneuver. It requires both hands—one to hold the

insect, the other to gently open its wings in search of the "family jewels"—
the conspicuous black dots that mark a male's pheromone sacs. This can
be a challenge in the field, when juggling butterfly net, pen, paper, and, of
course, a smartphone to document the procedure while gently but firmly
clasping a live butterfly.

Brower demonstrated how to ascertain a butterfly's sex with just one
hand. Just examine the butterfly's butt. If you see an obvious set of pincers
on the butterfly's rear end, it's a male. The deceptively delicate-looking black
protrusions are the same tiny claws that force females to mate and accept
their "nuptial gifts."

With half a dozen monarchs collected and safely stored in the cooler,
Brower also showed us how to tell if a female is carrying eggs. She has a "bead"
in her abdomen, which you can feel when gripping her gently.

We snapped a few photos of the drought-ravaged hillside, then set out for
Menard. Jenny had arranged access to a spectacular spring-fed pond rimmed
with dinosaur tracks and tall, white-blooming, well-hydrated frostweed. The
property, known back then as the Whispering Springs Ranch, felt like an
oasis. There we collected another thirty-four butterflies, all nectaring on the
late-season blooms. Quinn, Singleton, and I left the tour here, and Brower
and Kiphart continued on to Junction for a visit to the liatris fields at the
Native American Seed Company. Another forty butterflies were gathered for
Brower's drought study, and dozens more were tagged and sent on their way.

Flash forward sixteen months. One of the butterflies tagged that scorched
October day was recovered on the forest floor in Mexico. MJT 894, a male,
was found at El Rosario, the most visited and accessible of the five monarch
butterfly preserves open to the public. The news arrived via email on the
D-PLEX list on March 13, 2013. Diane Pruden, a retired human resources
executive who worked as a volunteer ambassador for Monarch Watch, had
recently returned from a visit to the sanctuaries. She shared the news that,
while unpacking from her trip, she had overlooked a hidden stash of recovered
tags that she had failed to report in previous emails to the monarch-centric
email list.

In Mexico, volunteer representatives from Monarch Watch are often
approached by locals and offered *etiquetas*, labels or "steekers," a Spanish
pronunciation of "stickers," to earn the $5 bounty paid per recovered tag.

The tags are sold back to Monarch Watch to assist in data collection for their monarch tag recovery database and to support the local economies of the remote mountain villages.

Pruden, one of Monarch Watch's superstar volunteers, has made the trek to Mexico several times, staying for weeks. She regales homebound butterfly fans on the D-PLEX list with descriptive updates of rewarding local people with bounties that constitute a line item of about $10,000 a year in the Monarch Watch budget.

Tag buyers are not official Monarch Watch representatives, but some, like Pruden, volunteer as Monarch conservation specialists. Others, concerned travelers with an interest in monarch butterfly migration conservation, volunteer their time and energy by offering to front the bounty costs to locals delivering tags. They later send in the samples and request reimbursements from bounties paid.

A volunteer tag bounty hunter requires a sense of adventure. One must pay for his or her own trip to Mexico, fend for oneself in Spanish, and navigate the remote mountain villages where people share the land with the butterflies.

During the winter, forest rangers, research associates, guides, and others visit the forest and look for dead monarchs with tags. The tags are removed and saved for the financial exchange with Pruden or other representatives. In the past, Chip Taylor personally went to collect the tags, but advancing age and the narco wars convinced him to stay home and send surrogates. After more than two decades of tag collection, the locals have come to anticipate the annual arrival of someone who will purchase tags. Sometimes the reps set up a table within the sanctuaries, and people line up. The tags are inspected for authenticity, then counted and recorded with the name of the collector and date retrieved. Each tag seller's name and tags are labeled separately. Pruden tallies the totals on her cell phone calculator to prove the math to the seller before paying up. She concludes with a thank-you speech in which she stresses the scientific aspect of the transaction as well as sincere gratitude.

Tag buyers typically photograph each tag set, organized by the collector's name. Then they email the photos directly to Monarch Watch, copying to their own email account to create a duplicate of the data on the web. Once returned to the states, the tags are packed and sent via UPS to Monarch Watch. There students and researchers review the tags and input the data into a database.

Monarch Watch counsels tag buyers to tread carefully in Mexico. They're advised to arrive unannounced, buy tags only in locations considered safe, and carry just enough cash needed for a given day. Restock cash, as needed, at an ATM in town.

In 2018, Monarch Watch again paid ninety pesos for each tag—about $5. In the past the organization had paid extra for tags still affixed to hind wings, as these were considered more valuable, perhaps because they contained actual butterfly DNA. But the organization changed the policy after determining that some tagged wings were fakes, attached after the fact to secure the tagging bonus. Circular blue-and-orange tags that are used in California are also sometimes recovered and submitted for payment. They receive the same rate.

Generally, Monarch Watch discourages taggers from sharing recovered tag data on the D-PLEX list or posting photos and other details on social media. The organization says posting the recoveries ad hoc makes for inefficiency, since they must continue to confirm and look up records in a less streamlined workflow. According to Monarch Watch, it's more efficient to turn over the entire list of tags to students to have them look them up systematically, but the organization has begun to work with social media groups in recent years.

About a week after Pruden's email, Singleton checked her logs and realized that nine of the tags Pruden listed were tagged during her October 7–13 stay in the Texas Hill Country in the fall of the historic 2011 Texas drought. MJT 894 was tagged on October 11, our day with Brower at Whispering Waters Ranch among the blooming frostweed at the natural spring that survived the drought.

Despite our success rate snagging monarchs for Brower's drought study, the butterfly population levels had dropped to their lowest levels since record keeping began in 1994. In March 2013, officials announced the population status for the 2012–13 season, the years that inherited the worst drought in Texas history. The entire migrating population occupied only 1.19 hectares. That represented almost a 60 percent drop from the prior year. In his annual monarch population status report on the Monarch Watch blog, Chip Taylor cited the usual monarch migration threats—habitat loss, genetically modified crops that allow for indiscriminate use of pesticides, and illegal logging in Mexico. He also cited "unusual weather"—the hottest, driest summer since record keeping began in 1895. "All in all, it was not a good year for monarchs," wrote Taylor.

How does one calculate the monarch butterfly population? The methodology that has been in place for decades is seriously flawed. Every winter, the World Wildlife Fund Mexico oversees the count. Technicians and scientists map the trees occupied by monarchs in each colony on foot, measure the perimeter of each, then multiply the number of hectares by a factor that over the years has ranged from 10 million to 50 million butterflies per hectare. The formula spits out a best-guess approximation of the number of butterflies overwintering in Mexico.

Lincoln Brower and other monarch scientists have expressed support for more accurate ways to measure the population since what's missing from the count is the density of the butterflies, which can vacillate substantially from tree to tree.[4] Drones and high-resolution photography could provide a more accurate census that counts not only the space but the density of the roosting butterflies.

Until the early 2000s, scientists estimated that each hectare of occupied forest represented about 10 million butterflies. Then a snowstorm in 2002 felled acres of forest. That allowed scientists to actually count the winter roosting butterflies that lay dead on the ground. Brower was there. He and teacher Dave Kust toured the destruction to come up with a number based on their firsthand observations. A single square meter contained 2,241 monarchs, suggesting 22.41 million monarchs per hectare, using the 10 million factor to calculate the population at the time.

But that didn't consider the many more surviving butterflies still roosting in the trees. On the same trip, Brower found even denser concentrations of monarchs roosting in the oyamels. As a result, a suggested factor of 50 million monarch butterflies per hectare became the accepted multiplier for estimating the population.

Until recently. In 2017 an article by Thogmartin and colleagues suggested that a "mean factor" of 21.1 million monarch butterflies per hectare would be more appropriate.[5] The researchers had evaluated various methods and locations of calculating the population at the monarch sanctuaries over several decades. The 21.1 million factor won out as the current one in use, although many charts and graphs still reference 50 million per hectare. Scientists prefer to simply quote the number of hectares and not reference a factor at all, since the calculations are so far from perfect.

One of the most frequently cited sources for communicating the declining population is the Monarch Watch population census bar graph. The orange-and-black-themed chart begins in 1994 and is updated annually with a new gold bar that reflects the numbers for the latest season. Journey North and Monarch Joint Venture also have charts reflecting monarch numbers for the eastern migrating population; the Xerces Society produces a chart for the western monarch population.

Seasons are calculated from March through December; thus, numbers announced in March of any given year actually reflect a population seeded the prior fall—the butterflies that made it to Mexico to overwinter and reproduce the following spring. Each season, then, actually begins in the fall of the prior year. For example, the 2017–18 season got its start with the survivors of the 2016 season; the historic low population announced in March 2013 started with butterflies that arrived in Mexico in the fall of 2012, which suffered the devastating consequences of the drought that began in 2011.

In March 2014 news broke that the monarch population had plummeted even farther from its historic low. World Wildlife Federation officials announced a monarch occupation of only 0.67 hectares. That's 1.65 acres, or 72,000 square feet—about 34 million butterflies if we use the generous 50 million per hectare factor, or only 14.1 million if we use the more conservative factor established more recently—a 90 percent decrease from the hundreds of millions in the late 1990s.

In his annual population census post in January 2014, Taylor blamed the usual suspects: pesticides, GMOs, and habitat loss. He also described at length the ramifications of climate variations: "It was the lower than normal temperatures in April that slowed development of immatures, and somewhat cooler May and early June that delayed the recolonization of the summer breeding areas." Basically, in the spring of 2014 the weather was just too cool during the breeding season.

Headlines trumpeted the end of the migration. "Monarch Butterflies Drop, Migration May Disappear," CBS News reported. The *New York Times* cast the dismal news in proper perspective: "The migrating population has become so small—perhaps 35 million, experts guess—that the prospects of its rebounding to levels seen even five years ago are diminishing. At worst, scientists said, a migration widely called one of the world's great natural spectacles is in danger of effectively vanishing."[6]

If there is an upside to the sad news, it is that the drought that defined the historic population lows unleashed a vibrant wave of monarch butterfly conservation advocacy. The 2014–15 numbers indicated a rebound population at 1.12 hectares, or close to three acres, but still low. In response to the bad news, the leaders of all three countries touched by the migration gathered in Toluca, Mexico, just seventy-five miles from the monarch butterfly ancestral roosting sites. *Los tres amigos*—U.S. president Barack Obama, Mexican president Enrique Peña Nieto, and Prime Minister Stephen Harper of Canada—publicly committed to work together to save the monarch butterfly migration.

In April 2014, first lady Michelle Obama planted the "first ever pollinator garden" at the White House.[7] The garden at 1600 Pennsylvania Avenue, which Mrs. Obama had installed as a kitchen garden in 2009, now included swamp milkweed, *Asclepias incarnata*, and butterfly weed, *A. tuberosa*. "Butterflies, bats, bees, birds: all of those, they get attracted to the gardens," the first lady explained to school children visiting the 1500-square-foot plot. "And then they go sprinkle life around so that food grows."[8]

In June, President Obama issued a presidential memorandum calling for the creation of the National Pollinator Health Strategy to promote the health of honey bees and other pollinators.[9] Presidential memorandums are one notch below executive orders in their heft. Some say the two are equivalent tactics, strong-arming the leaders of government agencies to take notice that the subject of the memo represents a high priority for the president. In the case of the National Pollinator Strategy, the directive was issued as a presidential memorandum, the same executive tool tapped by Obama to launch the controversial immigration policy known as the Deferred Action on Childhood Arrivals, or DACA.

By late August 2014, a group of conservationists had submitted the monarch butterfly for consideration as "threatened" under the Endangered Species Act. In a petition organized by the Xerces Society, the Center for Food Safety, the Center for Biological Diversity, and Lincoln Brower, the formal submission entered the queue as eligible for ESA protections under the secretary of the interior. The submission marked a turning point in awareness and appreciation for monarch butterflies.

By May 2015, Obama's National Pollinator Health Strategy was announced. The fifty-eight-page document outlined three ambitious goals for the United States: reduce honey bee colony losses by more than 15 percent within ten years; increase the migrating monarch butterfly population to 225 million (their historic average), with an occupation of six hectares (fifteen acres) by 2020; and restore or enhance 7 million acres of pollinator habitat over the next five years.[10]

Los tres amigos met again in June 2016, this time in Canada. The trio reconfirmed the Pan-American commitment to preserve the monarch butterfly migration.

This advocacy, progress, and attention to monarch butterflies and other pollinators following the migratory population bust can be directly attributed to the drought. But the angst provoked by the Endangered Species Act submission and President Obama's National Pollinator Health Strategy unleashed millions of dollars in government support for pollinator conservation. Nonprofit organizations and the private sector rose to the occasion of taking monarchs into the mainstream. By 2016 the monarch butterfly was pushed into the general public spotlight, well positioned as the poster child of climate change.

Three years later, eastern migrating monarchs made a dramatic rebound. The 2018 overwintering population in Mexico was announced in early 2019. It had vaulted 144 percent—the highest in a dozen years. Scientists and monarch watchers attributed the gains to a perfect storm of fantastic conditions: great weather, well-timed rains, and abundant host and nectar resources. Mother Nature seemed to be giving the eastern migratory population a reprieve.

But in California the story was much different. It seemed like a replay of our 2011. Drought plagued California for serial years and now the situation was dire. The Xerces Society, which has counted the California butterfly population since 1977, reported an 86 percent drop in their numbers. Dramatic headlines followed, predicting the end of the western migration and calling for more monarch conservation dollars.

"There's pretty good evidence that they're just not correlated," said Karen Oberhauser, when asked about the disparate story lines for monarchs east and west. "What goes on in the east is not connected to what goes on in the west because we have different weather patterns."[11]

In March 2019, the official U.S. Drought Monitor showed the entire state of California as drought free for the first time in eight years.[12] The weather reversal was attributed to atmospheric rivers, large vapor columns that move through the atmosphere and carry as much water as the average flow at the mouth of the Mississippi River. These atmospheric rivers cause winter storms that result in heavy rain and snowfall, which subsequently flood watersheds but also fill reservoirs and aquifers. News reports touted the official end of seven years of drought.

The wildflower "superbloom" that followed drew tens of thousands of visitors to the state to witness an epic display. Reports of a historic painted lady butterfly migration captured the imaginations of butterfly fans across the country. Billions of one of the most common butterflies on the planet entranced Southern Californians, who watched the masses of monarchs move down the California coast and shared the spectacle via social media.

Monarch aficionados couldn't help but ask, Will the California monarchs also make a comeback? In 2022 they did. The Xerces Society reported that their annual 2021 Thanksgiving count of 247,237 butterflies illustrated a considerable rebound from 2020's all-time low of less than two thousand. monarchs, which was preceded by two years of fewer than 30,000 monarchs.[13]

▼ ▼ ▼

CHAPTER 8

WELCOME TO THE MONARCHY

In his compelling book *Monarchs and Milkweed*, chemical ecologist Anurag Agrawal introduced the ecosystem surrounding *Danaus plexippus* and its host plant, *Asclepias* species. Agrawal detailed the battle between the organisms as they coevolved, casting them as biological enemies that have escalated their tactics over time. Confrontations occur on the leaves of milkweed plants. Monarchs consume them exclusively and voraciously; the milkweed responds by producing toxic chemicals. "The monarch exploits, and the milkweed defends," he writes, deducing that the creatures serve as royal representatives of all interacting species and labeling this system "the Monarchy.[1]

The turn of phrase is catchy. It accurately describes the ecosystem surrounding monarchs and their host plant. It also applies to another "Monarchy" that thrives far from the milkweed ecosystem and across the pond from the world's most famous royals. This passionate community spans America, crosses demographic and political lines, and includes professional scientists, citizen scientists, conservation professionals, and the butterfly-loving public.

This community of butterfly buffs and monarch aficionados closely monitors the monarch butterfly and all the related research and news. They are gardeners, conservationists, monarch taggers, and the rear-them-at-home fanatics who are simultaneously praised and castigated. One academic deemed them a "network of informed butterfly amateurs";[2] another, off the record, labeled some of them "people who simply have no capacity for thought."

This Monarchy includes everyone from casual butterfly buffs and devoted citizen scientists who've tagged thousands of monarchs to passionate, self-educated followers, some of whom engage incessantly in online discussion groups. Some track every academic article on milkweed or monarchs and monitor Google news alerts they've set up for "monarch butterfly," while many get their information from social media. Others raise hundreds of monarch butterflies in their kitchens and sign emails "Flutters." Some dress the part, wearing butterfly wings to monarch butterfly festivals, sporting headbands with antennae or donning orange-and-black themed T-shirts to announce their affiliation.

Those more casual in their interest care about gardening and landscaping best practices that increase pollinator habitat, but they don't have the time or inclination to follow monarch butterfly research or news. But some, a very small but vocal faction, serve as trolls, picking arguments, shunning reason and etiquette, creating drama and distraction. At times, spats have occurred in specialized social media groups and in email exchanges on the D-PLEX list, with name-calling and dismissive, impolite language a result.

Cultural anthropologist Columba González Duarte dug deep into this segment of monarch butterfly society in a 2018 study.[3] In it, she focused on the D-PLEX list, the listserv started and managed by Monarch Watch (see chapter 6). Her article presents the idea that cyberspace, and the D-PLEX in particular, have helped create a decentralized "rhizomatic" system in which power and influence are wielded not necessarily from a central entity or person but through a widespread and diverse system of participants.

Because of the D-PLEX and other internet-fueled platforms, the vast community of monarch butterfly followers has grown organically, with myriad entities interpreting and spreading their own versions of the monarch narrative via blogs, social media, websites, and email. In horticulture, rhizomes are creeping lateral stems that result in horizontal growth with adventitious roots. They serve as the primary stems of plants, mostly underground, but they possess the capacity to send shoots above the soil to absorb light, water, and air in order to thrive. The Monarchy exhibits this same capacity. It interacts consistently and intentionally with academia and the media in a three-legged stool of influence that determines the priorities, the narrative, and the future of the ecosystem it represents.

People who love monarch butterflies—and lots of them exist—give money to nonprofit organizations that support the insects' conservation. They click on the headlines that media companies assemble to draw readers to their websites and podcasts. They accept or reject the arguments laid forth by the pundits and professionals. They volunteer as citizen scientists and participate in crowdsourced data collection projects like the Monarch Larva Monitoring Project, Monarch Watch's annual tagging program, iNaturalist, and others. They plant—or not—the milkweed and nectar plants promoted for combatting the demise of all pollinators. And, intentionally or through misguidance, they sometimes promote misinformation.

The formally educated experts and paid professionals—mostly people with advanced degrees in science who focus on monarch butterfly and milkweed research—stake their credibility and voice on how many articles they've published, how proficient their public relations and social media skills, or how much money they've raised on behalf of their institutions. Their willingness to engage with reporters on deadline, no matter how inconvenient, also impacts the amount of influence they carry.

News coverage is ample because the media understand how monarch butterfly news draws readers. The National Wildlife Federation's 2015 Mayors' Monarch Pledge campaign has made monarch butterflies "local news" in hundreds of municipalities across the country by working to get those cities to commit to pollinator conservation. Media are obliged to cover it. The seasonal predictability of the monarch butterfly migration and its journey through three countries also creates timely stories regarding the "state of the union" of the butterflies. Will the population be up or down this year? When can we see the international travelers in our corner of the world?

Thanks to this triumvirate—the media/nonprofit communities, academia, and the Monarchy—general interest in monarchs is high. Traditional media and online news sites know that monarch butterfly news draws readers. And they are not above crafting dramatic headlines to fuel this effect. Nor are nonprofit organizations shy about tapping into perceived fears of demise to maximize fundraising efforts.

Having participated in the management of a general news website for years as well as a niche website devoted to pollinators, I have witnessed the effect of monarch butterfly news coverage on readers. Readership stats on stories with "monarch butterfly" in the headline are generally higher than other nature news. And the seasonality of the annual migration makes for at least three predictable news cycles per year.

In the spring, when monarch butterflies leave their roosting sites in Mexico, the hopeful—or dreadful—"here they come!" stories abound:

▼ "Monarch Butterflies Face 'Quasi-Extinction'—But Hope Is on the Wing," NBC News, April 2016
▼ "Expect 300 Million Monarch Butterflies in Texas This Season," *Houston Chronicle*, March 2019
▼ "Endangered Monarch Butterflies at Risk as Climate Changes," *Washington Post*, March 2023

At some time either before or thereafter, a population status update from Mexico occurs to relay how many hectares/acres of forest the monarch butterfly roosts occupied during the previous season. This official report, organized by CONANP, usually arrives in February or March but has been released as late as May in recent years. Another round of news stories results:

▼ "Monarch Butterflies Drop, Migration May Disappear," CBS News, January 2014
▼ "Butterflies Abound in Mountainous Mexican Winter Habitat," Associated Press, March 2019
▼ "The Monarch Is Falling Victim to a Real-Life Butterfly Effect," Vox, December 2021

Then in the fall, after the multigenerational migration moves south from Canada through the United States to Mexico, another seasonal update occurs:

▼ "In Mexico, Hopes of Uptick in Dwindling Monarch Butterfly Migration," *Christian Science Monitor*, December 2015
▼ "We're Losing Monarchs Fast—Here's Why," *National Geographic*, December 2018
▼ "Monarch Butterflies May Be Thriving after Years of Decline: Is It a Comeback?" *Guardian*, November 2021

If you notice a bit of whipsawed drama in the headlines above, you are not alone. Members of the Monarchy notice it, too, and constantly seek understandable explanations. Insect populations are famously volatile, booming one season then busting the next, depending on weather and other conditions.

The same is true for the Americas' favorite insect, but the public doesn't always grasp the concept.

Some of the PhDs who study monarchs and the paid professionals at non-profits devoted to their conservation try to make the science understandable to the Monarchy. Some of these professionals truly serve as what I like to call "scientist citizens," making every effort to educate the community in a helpful and approachable way. But a few often talk down to the Monarchy in a condescending tone with a "bless your heart" attitude. It doesn't help that various studies declaring completely different theories about why the butterfly is or is not in decline compete with each other for votes as the accepted narrative.

For example, migration studies expert Andy Davis reviewed more than two decades of research accounting for the monarch butterfly population in the United States. Appearing in *Global Change Biology* in June 2022, his article reviewed twenty different previous studies based on twenty-five years of observational data from the North American Butterfly Association.[4] It's conclusion: monarch butterfly numbers are *not* declining.

On his Science of Monarch Butterfly blog, Davis summarizes the study, declaring that after reviewing the trends from more than four hundred monitoring sites "no overall decline in numbers of monarchs [was] seen in the past 25 years, going back to the mid 1990s." In fact, Davis determined an overall positive trend of 1.3 percent per year—about a 30 percent increase in monarchs over twenty-five years. In conclusion, writes Davis, "Monarchs do not really need our help, they just need to be left alone." His article reinforced what became known as the "migration mortality hypothesis."

Davis made the case that the number of monarchs reaching the overwintering sites is declining but the overall population is not; thus, increasing monarch mortality during the migration over the previous two decades is what accounts for the decline in monarch numbers. As he told the *New York Times*, this is not a production problem.[5] "We don't have fewer monarchs. We have fewer monarchs reaching the wintering colonies."

Monarch Watch founder Chip Taylor, a big proponent of the generally accepted "milkweed limitation theory"—that a lack of available host plant in the summer breeding grounds is the primary explanation for monarch decline—wasn't having it. Taylor released his own study, which completely contradicted Davis.[6]

Taylor's study evaluated tagging data from 1998 to 2015. His research found that a correlation indeed exists between the size of the late summer breeding population in the Midwest and the monarchs' overwintering population. Migration success was not correlated with the size of the overwintering population size, and migration success did not decrease during this period, the study found. Rather, migration success was correlated with "the level of greenness of the area in the southern U.S. used for nectar" by migrating butterflies, providing the fuel for them to build up their fats and survive the flight and the winter.

Thus, according to Taylor, the primary factor in annual variations in the overwintering population is summer population size, with migration success being a minor determinant. "Increasing milkweed habitat, which has the potential of increasing the summer monarch population, is the conservation measure that will have the greatest impact," Taylor wrote.

What is the Monarchy to make of the fact that two of the most vocal monarch scientists in the country face off with competing narratives regarding the most important determinants of a healthy migrating monarch butterfly population? One, who specializes in the study of migration, posits that the migration itself is the most debilitating factor. The other, who lives in the heart of the midwestern summer breeding grounds where milkweed once flourished but is now less available, contends that lack of milkweed habitat is the problem.[7]

Confirmation bias, the tendency to find what you're looking for, is pervasive in all corners of life—even science. We're all guilty of it. My husband loves Indigenous artifacts. When we walk at the ranch, my husband Bob Rivard looks to the ground for arrowheads and lithic tools left by Comanches and the other Indigenous peoples who occupied our stretch of the Llano River. Several showcases of Perdiz points, atlatl spearheads, and other artifacts sit on our counter, a testament to his obsession.

I favor butterflies and insects and the plants on which they rest, nectar, and lay eggs. On those same walks, I'm looking at the bushes and flowers that hug the earth. I stop to look under a tender leaf or aim my iPhone camera at a bloom with unidentified insect visitors, to be researched later. My head is in the forbs, and I have learned much from them.

When birder friends visit, they keep their eyes trained on the trees and the skies, in search of avian life. Binoculars in hand, they set their sites on the pecans, persimmon, scrub oaks, and bird feeders near our house—a vista generally twenty feet or higher than Bob's or mine. Their reward: multiple additions to their birding life lists.

▼ ▼ ▼

CHAPTER 9

BUTTERFLIES AS BAUBLES

For about six months at the end of 2010, I thought I wanted to become a professional butterfly breeder. The idea held great appeal in the context of the downward spiral of the media business in which Bob and I had invested our professional lives. Cost cutting and downsizing prevailed at the newspaper where Bob worked, and it was clear that the same scary tide would soon hit Business Wire, the commercial wire service where I spent fourteen-plus years. The old fashioned "wire" was entirely dependent on traditional media to distribute press releases for corporate clients. The internet made possible what was previously impossible—the widespread sharing of information through social media. It also left few, if any, industries unscathed.

After my early tagging experiences and deep dive into all things *Danaus plexippus*, I found myself reading *The Family Butterfly Book* by Rick Mikula. It inspired me and made me think I might like to raise butterfly livestock professionally someday. Mikula, a former machinist, is credited—or cursed, depending on your perspective—with launching the controversial industry of commercial butterfly breeding. The bearded butterfly evangelist from Pennsylvania and his wife Claudia came up with the idea of releasing live butterflies at weddings and funerals back in the 1990s. In person and in his writing, Mikula demonstrates an unmitigated love of butterflies. He's raised every common North American species and many exotics for decades. In the early 2000s, he translated his fascination and firsthand experience into a traveling road show called the "Spread Your Wings and Fly Seminar."

His touring workshop taught the basics of becoming a commercial butterfly breeder and moved up and down the East Coast in the early years of the twenty-first century. "If you decide to raise and sell butterflies, be prepared to be sold out at least a year in advance, every year," teased the promotion for the six-hour course. For $245, participants were schooled in the business of butterfly breeding. Mikula promoted the seminars on his website: "From permits to protozoa, marketing to mailing, and raising to releasing, everything will be covered in this extensive day-long program."[1] Lunch included.

Mikula converted his niche, hands-on knowledge into *The Family Butterfly Book*. The thin workbook was an esoteric resource when it was published in 2000. I discovered it a decade later, but its appeal persisted. It captures the infectious, fun-loving spirit of a man who cannot resist sharing his profound affection for and interest in order Lepidoptera.

Written in a "silly uncle" style and packed with esoteric facts ("some butterflies really like kitty cat urine!"), Mikula's book captivates the whole family. He makes the complexities of metamorphosis and plant-insect interactions compelling and understandable. One section offers a guide to the forty most common backyard butterflies. Details of their range and host plants are accompanied by color illustrations of each species in all their life stages—egg, caterpillar, chrysalis, and adult. Chip Taylor of Monarch Watch wrote the book's foreword, providing scientific credibility and, for a time, earning 1 percent of the book's profits for Monarch Watch. *The Family Butterfly Book* serves as the perfect butterfly primer, offering science, fun facts, practical knowledge, and hands-on activities in one easy-to-read package.

Chapter 4 intrigued me: "Starting Your Own Butterfly Farm." Here, Mikula details how to raise butterflies in captivity. Sections describe the best way to build a caterpillar cage, how to use bleach for "good housekeeping," and tips for establishing a "maternity ward"—a clean, safe nectar haven for freshly hatched adults. He also discusses the importance of not crowding, even how to "play matchmaker" and hand-pair (captive-breed) your butterflies. As I ripped through the book, I was titillated by the prospect. Breed butterflies for a living? I *loved* this idea.

Soon I was searching the website of the International Butterfly Breeders Association, a trade association of professional butterfly breeders from all over the world that Mikula referenced and helped start in 1998. I paid the $160 membership dues and gained instant access to the IBBA email list as well as a trove of members-only insights housed behind a firewall. Most surprising was how the IBBA email list worked as a virtual commodities exchange. Here, in the virtual stockyards of the internet, breeders bought, sold, and traded butterflies in all their stages. "Anyone have Giant or Black swallowtail pupae for sale?" "I need 100 PL (Painted Ladies) shipped ASAP." "We have extra monarchs available. Send me an email if we can be of service."

Such requests and solicitations fueled the IBBA listserv—and yes, at the time it was an old-school email list. Members of the tight community, competitors but also collaborators, helped each other with advice and guidance and covered for each other when one encountered an unexpected inventory shortfall. This butterfly livestock marketplace seemed to have the potential to gird a small, start-up business. The knowledge that you could lean on your fellow breeders for advice, guidance, and inventory gaps created a sense of community and solidarity. What was abundantly clear: breeding butterflies is a seasonal and volatile business. Surpluses and shortfalls come with the territory.

Excess butterflies are easily handled. Just offer them on the listserv. No takers? Release them to the wind. Shortfalls can be more challenging. For large orders tied to specific events, the breeder had typically contracted with a buyer and received a deposit of half the order's total cost, at a set price. Then a catastrophe hit his or her operation. A lack of host plant, disease, predator invasion, bad planning, bad weather, or simply bad luck could be to blame. Yet the breeder *must* come through with live, healthy butterflies, on deadline, for that much anticipated and meticulously planned wedding, funeral, festival, museum exhibit, or educational event.

In such situations, the breeder would issue an email plea to the group, often driving the price per butterfly substantially beyond its quoted cost, to deliver the livestock to the customer as promised. Sometimes a breeder would request that butterflies be drop-shipped directly from the subcontracted breeder to the customer. Arrangements could even be made for the shipper to use the branding and return address of the breeder who originated the order, so that consistency and client relationships could be maintained. These practices continue to this day.

After spending a few weeks reading the voluminous email strings and gaining an understanding of the IBBA's role in the butterfly breeding business, I learned that the association would stage its annual convention in Las Vegas that fall. I blocked out some vacation time at work and signed up for the conference, which took place at the unlikely Circus Circus Hotel.

My introduction to my future butterfly breeding colleagues debuted on the Las Vegas strip, where Lucky the Laughing Clown, the neon-lit icon of the gambling house that Jimmy Hoffa built, straddled the parking lot of one of Vegas's oldest casinos. The aging, garish joker waved and pointed to the casino entrance with one hand, holding a twirling, flashing whirligig in the other. As best I could tell, the strongest connection between the Las Vegas strip and the butterfly breeding business was a shared economic interest in the wedding industry.

Somehow, I convinced my butterfly-tagging buddy Veronica Prida to come with me on this trip. Veronica is a fellow gardener, former neighbor, nature photographer, godmother to our sons, and dear, loyal friend. Born in Mexico and educated in San Antonio, she's also a successful designer, artist, and businesswoman. We share an adventurous spirit and travel well together, as in the week we had recently spent in southern Mexico where she was working on a project.

On that trip, Veronica had arranged an outing that allowed me the unforgettable experience of chasing blue morphos in their native jungle habitat outside Huatulco, Oaxaca. By then I had made myself a rule to stop running blindly after butterflies with my net. Too often I had done so, often on the Llano River, chasing monarchs in the fall. Sometimes I would trip on a rock or slip on wet limestone in the middle of nowhere with the closest hospital hours away.

But the sight of a blue morpho, one of the largest and most beautiful butterflies on the planet, its iridescent aqua wings flashing languidly against the tropical canopy, was irresistible. The butterflies were tracking a dirt road in the coffee-growing fincas of Mexico's Sierra Madre. Running full speed ahead while looking up, I chased a blue morpho for about five hundred feet before tripping on a fallen branch. Luckily, I caught myself.

Veronica and I arrived late to the IBBA reception, finding our way through the crowded casino where staff dressed like Elvis and other celebrities dealt blackjack and Texas hold'em to willing players. Middle-aged ladies crowded around the "one-armed bandits," feeding quarters into the slot machines and sipping watery gin-and-tonics from short plastic cups.

The IBBA welcome reception unfolded in a modest hotel room. A collection of round tables with white linen tablecloths and a makeshift bar filled the space. Breeders from Costa Rica, Canada, and the Pacific Northwest traded hugs and handshakes with their peers from Florida, Texas, Pennsylvania, and elsewhere. Everyone in the small crowd seemed to know each other. Keynote speaker, lepidopterist, and invertebrate conservationist/activist Robert Michael Pyle mingled with butterfly breeders from around the globe. His book, *Mariposa Road*, had just been released, and he was delivering the keynote speech. Veronica and I introduced ourselves—we were with the Texas Butterfly Ranch, and we were considering starting a butterfly breeding business.

An intense passion for butterflies and the responsible practice of rearing healthy, beautiful lepidopterans seemed to bind this odd, quirky tribe. They were warm and welcoming. Everyone offered encouragement, wisdom, and advice. I learned much and made contacts I would lean on as I developed my plan to hatch the Texas Butterfly Ranch.

Veronica and I attended sessions on how to raise swallowtails, assemble exhibits, and market to the wedding industry. We heard from Wayne Wehling of the USDA, via Skype. At the time, Wehling oversaw the regulatory process surrounding the interstate transport and trade of butterflies.[2] Only nine species are permitted to cross state lines. Professional breeders must apply for a USDA permit for each of the fifty states to ship livestock outside their base. Breeding operations, like nurseries, are also subject to inspection.

I didn't know it at the time, but keynote speaker Pyle was one of the most vocal critics of the interstate transport of butterflies and live butterfly releases on earth. That he was invited to speak to the IBBA was a remarkable strategic move by the organization to educate, and possibly disarm, one of its boldest detractors.

As the founder of the Xerces Society for Invertebrate Conservation in Portland, Oregon, the Yale-educated entomologist and Guggenheim fellow had recently ripped the commercial butterfly breeding industry in an essay published in the Lepidopterists' Society newsletter. In the summer of 2010, Pyle characterized the practice of commercially breeding butterflies in one area of the country and releasing them in another as "a thoroughly unhelpful and disruptive practice." He made his case in a scathing 3,300-word essay titled "Under Their Own Steam: The Biogeographical Case against Butterfly Releases."[3]

Pyle's expressed concern was that the release of commercially bred butterflies away from their birthplace would mess with scientists' data. This
would be incredibly inconvenient, he pointed out in his essay: "It is easy to
see how releases could perturb butterfly monitoring transects, annual butterfly counts, our society's Season's Summary and many other measures of
presence and absence."

Almost a decade later, Pyle expressed adamant opposition to the shipping,
transport, and release of butterflies across state lines. In a February 2019
interview, he said that his concerns had been misconstrued for decades.[4] In
fact, he professed to not be opposed to commercial breeding at all, and to have
never been so. He had helped set up breeding shops for exotic butterflies in
Papua New Guinea decades earlier. It's what people did with the butterflies
afterward that bothered him, he stressed poignantly. His primary concern
all along: the butterflies' biogeographical aspect—that is, their geographic
distribution.

For the conservation of butterflies to be successful, Pyle believes, we have
to know their "normal" movement patterns—where they are at what time of
year and how that changes over time. "It's utterly essential with monarchs
and other butterflies to know their natural movements in nature, what their
natural whereabouts are," he said. When creatures are moved around through
human agency to locations they didn't find on their own steam, "nature is
in danger of being seriously disrupted." That's why Pyle believes regulations
should exist to forbid the breeding and shipping of butterflies beyond the
provenance of their natural origin.

In 2009, Pyle and the Xerces Society worked to get the USDA's Animal
Plant and Health Inspection Service to pass a regulation to halt the import
of European-bred bumble bees, which were being used in commercial greenhouses to pollinate tomatoes. The imported bees were introducing new diseases to the domestic bumblebee population, including the Franklin's and
western bumblebee, as well as the rusty-patched and yellow-banded bumblebee in the East.[5] In the Lepidopterists' Society essay, Pyle likened commercial
breeding and the staging of butterfly releases to playing "animal chess" and
dubbed commercially reared livestock "virtual lab clones masking as butterflies" "Do we really want our butterfly fauna to suffer the same tossed salad
treatment that our native flora has withstood?" he wrote.

Festivals, funerals, weddings, and educational exhibits are important markets for commercially reared butterflies. Photo by Matt Buikema, courtesy of Pearl, San Antonio.

Pyle closed his piece by alerting his fellow lepidopterists that he was venturing "willingly into the lion's maw this fall" at the IBBA conference in the interest of finding common ground. And as he vilified the butterfly breeding industry in the professional butterfly scientists' newsletter, he softened his tone and slightly lowered the volume of his condescension when describing the association membership. Most IBBA members "really love butterflies, as we do," he wrote. "But most did not begin as collectors, and therefore lack our sense of excitement based on butterflies' natural whereabouts. . . . They simply fail to understand the real concerns their activities raise."

Veronica and I did not attend Pyle's talk. We had tickets to see the Beatles' Cirque du Soleil LOVE show at the Mirage, and it was spectacular. Pyle would visit San Antonio the following week on his book tour, and I planned to attend his talk at our local independent bookstore. Those who heard Pyle speak relayed that he was perfectly civil. He expressed his reservations about butterfly releases in kind language. He read from his book and signed copies for those willing to purchase it.

By the time I quit my job and moved back to San Antonio in late 2011, I was busy securing my permits to raise and ship butterflies across state lines when John C. Abbott called me in San Antonio. Curator of Entomology at the University of Texas at Austin at the time, Abbott wondered if I might be able to supply live butterflies for the university's first, and what turned out to be only, Insecta Fiesta, a celebration of invertebrates. The insect guidebook author and renowned expert on dragonflies had heard about me through my friends at the Austin Butterfly Forum, where I had been attending meetings. He was aware that I was starting a butterfly breeding business. Abbott was a big fan of the famous Bug Bowl at Purdue University, an annual two-day event that draws 30,000 people to the Indiana campus. In addition to a cricket spitting contest, cockroach races, art painted by caterpillars, and an entomophagy tent in which insect-inspired offerings like banana-ant bread are made available to a wary public, Abbott wanted to set up a live butterfly house. Could I supply five hundred live butterflies for the inaugural event, which would take place at the Brackenridge Field Labs where I had taken Botany for Gardeners as a freshman at UT?

I assured Abbott I could, even though I was living in an apartment with no yard or greenhouse and my only experience raising butterflies was rearing a few swallowtails and monarchs in pop-up cages. But I knew that by working with my new friends at the IBBA I could tap the butterfly commodity exchange and put out a bid for a mix of monarchs, painted ladies, Gulf fritillaries, and other native Texas butterflies.

I followed the protocols of the IBBA list and sent out an email request two months before the insects were needed:

Hey there IBBA folks,

Well, I got my first order—it's for an Insecta Fiesta to be held in Austin, TX on April 21. Yeah! I need 500 butterflies, all native Texas species, to be utilized in a fly house for a one-day event. Since I'm in no position to raise these myself at the moment, I'll need you all to fill the order. Please email me directly if you're in a position to provide. Thanks, always, for all your support and help.

—Monika

Five hundred butterflies is a hefty order for an individual butterfly breeder. Generating high-quality, clean, healthy stock on deadline is often a group effort, with various breeders pitching in. Several breeders offered to provide painted ladies and perhaps monarchs, but seemingly no one could provide the mix I was seeking. Finally, I ended up working with Connie Hodsdon of Flutterby Gardens, Bradenton, Florida. Hodsdon, a petite, no-nonsense former insurance agency owner, touted her attention to butterfly detail on her email signature: "Connie. Owner of the cleanest butterfly flight houses in the world."

Hodsdon was extremely generous and reassuring while working with me, taking charge of the livestock and encouraging me to hustle more sales. "You have a great weekend, too, and think of your next client. I'll handle this," she wrote me when I sought her advice on how to deal with other breeders' mixed messages on supplying the livestock I needed. She directed me to my first microscope, which I would use for OE testing, advised me on packing livestock ("always pack and ship them out the last thing of the day so the ice doesn't melt"), and provided guidance on basic business questions, such as whether butterflies are taxable. Yes, they are. In exchange, I helped Connie start the social media outreach for her business. We made a good team.

The Insecta Fiesta took place on April 22, 2012. Thousands of visitors arrived at Lady Bird Lake for the free event. The butterfly house was one of the most popular features, and we even witnessed a monarch butterfly eclose, or drop from its chrysalis, in front of visitors. Children and their parents watched in amazement. At the end of the day, we opened the doors and roof and the butterflies flew off. Cost for the spectacle: $3,500.

The Insecta Fiesta made crystal clear what it would take to succeed as a professional breeder: a religious embrace of systematic protocols and a commitment to doing the same thing, the same way, methodically and repeatedly for every order placed. That would be the only way to produce live, healthy butterflies. This realization contributed to my decision to stick to butterfly evangelism and advocacy. My strength does not lie in doing the same thing, the same way, every day. The Texas Butterfly Ranch would remain a pollinator advocacy site, and if I ever needed live butterflies I knew where to find them.

▼ ▼ ▼

CHAPTER 10

ARE YOU CATHY AGUADO?

With my professional butterfly breeding aspirations behind me, the drought stretched into 2012 and I continued posting stories on the Texas Butterfly Ranch website. One spring morning I was sipping my second cup of coffee on a sunny Saturday back in San Antonio, about a month after the Insecta Fiesta, when I discovered a new email in my inbox, time-stamped 6 A.M., May 24, 2012:

> *I am the only living member of the team who discovered the monarch butterfly overwintering sanctuaries in Mexico in 1975. The discovery was published by National Geographic Magazine in August 1976. My picture is on the cover. I was referred to as Cathy back then. I have been here in Austin living a quiet life and I am interested in participating in your Austin Butterfly Forum.*
>
> *—Best regards, Catalina*

I gasped audibly. Catalina's cover girl moment on the cover of *National Geographic* and Fred Urquhart's personal narrative of fulfilling his lifelong quest to find the monarch butterflies overwintering sites jumped to mind. The "NatGeo story," as it's known in monarch butterfly circles, literally put monarch butterflies on the coffee tables of middle America. Anyone who pursued an interest in monarch butterflies had read the article and been captivated by the shot of the joyous Mexicana on the cover, grinning broadly, surrounded by millions of monarch butterflies.

Really? The woman busting through the magical wall of monarchs lived and worked in Austin? Just seventy-five miles up the road? Why had we never heard from her? And how many times had I looked at that cover photo and wondered, Who is she? What was she thinking? How did it happen? She's so lucky.

Her email—a fluke of mistaken identity—arrived in response to a post I'd written the previous week about a meeting of the Austin Butterfly Forum. At their April 2012 meeting, the butterfly conservation and education organization had invited its members to gather and share tips on raising butterflies at home. The Texas Butterfly Ranch covered the event.

Catalina had read the story early in the morning. Her limited web savvy misled her to believe she was communicating with the Austin Butterfly Forum—when in fact she was posting a comment to the Texas Butterfly Ranch.

I responded immediately: "Are you Cathy Aguado? Wow, just saw your email. . . . I always wondered about you, and what it must have felt like to first walk in the roosting forests. Would love to interview you for an article some time. Are you up for it?"

Catalina Aguado of Morelia, Michoacán, was known as "Cathy" for decades—even though she didn't care for the *gringa* version of her name. Her then-husband Ken Brugger called her by the less-lyrical American version, probably because it was easier for him to pronounce. In the NatGeo story, Urquhart mentions a "bright and delightful Mexican" named Cathy Brugger. English-speaking chroniclers of monarch butterfly history adopted this moniker. In the annals of monarch butterfly history, Catalina became known as "Cathy." "Yes, I am Cathy but have returned to use my complete name, Catalina. . . . Sure, it would be nice to talk about the monarch butterfly discovery. I stayed under the radar for so long."

Her word choice gave me pause, suggesting the need to evade detection, harm, or destruction. Literally, a survival technique.

We connected by phone. When Catalina answered, "CAHT-ah-LEE-Nah," I felt an instant bond. My name, too, sounds lovelier when spoken in its given language—MOH-nee-kah in German versus MAH-nih-kuh in English.

We agreed to meet the following week in Austin. Sitting in a coffee shop near her house, Catalina wore her silky dark hair pulled back into a ponytail like the one in the famous *National Geographic* photo. The vibrant young Michoacána that we remember was present, but caution replaced the joyous smile she wore that day on the mountain when the photo was snapped. She exuded vigilance and spoke softly.

Over a glass of iced tea and the course of the next month, Catalina cautiously unraveled what has come to be known as the monarchs' "discovery story." Often she requested I delete details that she inadvertently let slip in moments of conversational candor. She was especially sensitive to protecting her son's privacy. And she told me repeatedly, "The truth of my story is important." I did not disagree.

Born in 1949 on a ranch in the remote mountains of El Salto, which means "the jump" in Spanish, Catalina grew up outside the Michoacán state capital of Morelia. She came into a world where expectations of her were to marry young and "have children, until the end of my reproductive life in a 'survivor of the fittest' environment."[1]

But Catalina Aguado rejected that path. She married a dashing gringo from Kenosha, Wisconsin, thirty-one years her senior. They made a life together and shared many adventures—a partnership, a child, travels, and tumult. Their shared accomplishments included unlocking a venerable mystery of nature in the oyamel forest of Cerro Pelón, just 120 miles from Catalina's birthplace. It rocked the world of lepidoptery. And Catalina's role, instrumental to its occurrence, had largely been overlooked. The events she would share with me came to reveal a much more complex story, one that once again would rock the Monarchy.

Over the course of many phone calls and emails and several in-person meetings, a friendship grew. Catalina shared the story of engaging the hesitant campesino with the horse—"so we wouldn't be alone." She recalled the strenuous trek through the forest to the monarchs' winter hideaway. She remembered the shortcut, the concerns about coral snakes, the scary sensation of wading in deep forest mulch, the towering oyamel firs, and the silent magic of the butterflies clinging to the tree trunks in the chill mountain air. The unpacking of the tripod, taking the photos, waiting for Ken to catch up, yelling, "We see them! We see them!"

Her story, as dramatic as her husband's, had been tucked away for decades. We told it together for the first time, from her perspective, at the Texas Butterfly Ranch: "Founder of the Monarch Butterfly Roosting Sites in Mexico Lives a Quiet Life in Austin Texas." Catalina even retrieved photos of the historic day from her private collection for use in the article.[2] I suggested we add a copyright line to the photos, which we did. The story was published on July 12, 2012.

▼

Earlier that year, SK Films tried to locate Cathy Brugger to serve as a consultant on a movie project, *The Flight of the Butterflies*. The award-winning, Toronto-based production company was bringing the monarch butterfly discovery story to 3-D IMAX screens across the continent. The forty-minute docudrama would focus on Fred and Nora Urquhart and their decades-long quest to find the overwintering sites. Ken and Catalina played supporting roles. The filmmakers attempted to track Catalina down, but it was practically impossible. Almost everything in her life had changed since the historic day in 1975 when she and her husband had scaled Cerro Pelón.

Catalina and Ken relocated to Austin and had a son. Catalina earned a college degree in social work from the University of Texas at Austin. And after eighteen years of marriage, Catalina and Ken had divorced. She's been married to George D. Trail, a fellow social worker, since 1995 and had taken his last name.

By the time the producers sought her assistance on *Flight of the Butterflies*, Catalina had been absent from the monarch butterfly scene for more than

Catalina Trail and William "Bill" Calvert at an Austin coffee shop in June 2012. Photo by Nicolas Rivard.

thirty-five years. A lifetime. The rivalries for recognition that accompanied the discovery story greatly dismayed Catalina when they unfolded during the late 1970s. They drove her away. Lincoln Brower and Fred Urquhart's "unpleasant incident in the forest" was just one unnecessary drama. Other incidents had created gratuitous stress over the years.

In one instance, before Urquhart's first visit to Mexico and the official announcement in *National Geographic*, Ken Brugger extended an ill-advised invitation to friends from Texas to come and see the roosting sites. He specifically instructed them: "Keep it to yourself." They didn't. The duo tried to take credit as "co-discoverers" of the sites, even threatening *National Geographic* with legal action, according to Catalina and others.

Carlos Gottfried, a Mexican businessman who worked with Brower and Urquhart in the 1990s to establish the monarch butterfly sanctuaries, said the potential for a nefarious lawsuit contributed to the curious twenty-month delay in publishing the story in *National Geographic*. "Everybody was pissed off!" said Gottfried.[3] Catalina recalled that *National Geographic* editor Gilbert Grosvenor was annoyed with her husband for exercising such poor judgment. The issue was eventually resolved, but bad feelings lingered.

Ken Brugger frequently told inaccurate stories about what had happened at Cerro Pelón. That bothered Catalina. The false accounts by her husband published in Sue Halpern's 2001 must-read book for monarch butterfly fans, *Four Wings and a Prayer*, especially grated on her. Ken was seventy-one by then, showing signs of dementia and often misrepresenting reality and embellishing the narrative. "I told him it was wrong, and he said it didn't matter."

Ken Brugger had told Halpern that he and Catalina traveled through lots of dangerous territory in their search. "People threatened to shoot us. They told us that Zapata had hidden some gold up there and they thought we were looking for that," he said. Urquhart expanded the narrative, describing a scene in his *National Geographic* story whereby "Mexican woodcutters, prodding laden donkeys, had seen swarming butterflies and had helped point the way" to the roosting site.

Catalina, who was there, speaks Spanish, and grew up in the region, told a different story. She remembered Winnebago breakdowns, arduous climbs up difficult mountain trails, hundreds of exchanges with locals, but not the drama described by Ken. "That's not the way it happened, and Ken never corrected that," said Catalina.

In 2009, Catalina and George Trail decided to downsize. The move from a two-story home to one without stairs required a substantial edit of their belongings. Catalina had been accumulating collectibles for twenty years, hoping to start a second-hand shop someday. Years of stockpiling "pretty things" had turned her garage into a packed warehouse.

To prepare for a mammoth garage sale, she purged hundreds of books, furniture, knick-knacks, and other belongings. She also shredded a stockpile of documents, including copious notes taken during her time chasing monarchs. Among the destroyed papers was a carbon copy of the letter Ken wrote to Fred Urquhart in which he offered to volunteer. Catalina had typed it and all their field notes on her trusty Smith Corona typewriter. For decades she had banished these monarch papers to the dark recesses of her garage. The renowned set of oxygen tanks the Urquharts' used in their high-altitude climb to see the roosting sites for the first time was stored there, too. Years later Catalina recalled the petite green tanks as "cute" and described their handy packaging as an "old fashioned little suitcase with a lock and everything." But the time had arrived to let all of the artifacts go.

She shredded much of the evidence of her years chasing monarchs. It served as a cleansing for her. She felt better, detoxed. Catalina had tucked away those years, something many of us find difficult to fathom. Who knew that three years later monarch butterflies and their natural history would become a subject of mainstream fascination? Not Catalina.

Interviewing her for the first time, I asked if she had ever told her son monarch butterfly bedtime stories about the time she and daddy chanced upon the mountain hideaway where millions of butterflies spent the winter. She shook her head, no.

"I forgot about it," she said. "Until 2012 comes to hit me in the face. And SK films finds me."

▼

The producers of SK Films exhausted all avenues trying to track Catalina down. They tapped the usual channels—Google, public records, phone books. They hired a private investigator. They reached out to the Monarchy—the scientists, amateur naturalists, and butterfly fans who follow the D-PLEX list. Don Davis, one of the original members of the Urquharts' Insect Migration Studies Association and an avid D-PLEX contributor, helped make the connection.

SK Films invited her to be a consultant on *Flight of the Butterflies*. As the only living member of the discovery party, she was uniquely qualified to provide a personal, truth-based account of that day. She told me multiple times over the years that the producers didn't follow the truth to the letter—it interfered with their narrative.

In March 2012, Catalina took time off work to fly to Mexico to provide guidance on the film—personal details of her and Ken's travels, and their time with the Urquharts. She even offered guidance to rising Mexican film star and future James Bond girl Stephanie Seligman, who played her in the movie.[4]

In September she again made a whirlwind trip, this time to join the SK film crew in Washington, D.C., for the opening. Don Davis was there, too. He shared an exuberant email with readers of the D-PLEX. Touting the film, he noted that as the lights dimmed at the Smithsonian Museum of Natural History "a gentle lady" found a seat next to him. It was Catalina Aguado Trail, formerly known as Cathy Brugger. "We finally met!" wrote Davis, adding that as he traversed the mall on his way to the screening a monarch butterfly crossed his path. "A good omen!"

In March 2013, Catalina participated in the Mexico City opening of the film. There the State of Mexico presented her with the 2012 "Jose Maria Luis Mora" Gold Medal Award for "relevant and eminent merits and conduct of notable service to humanity, Mexico, and The State of Mexico."

The gush of attention and publicity must have reignited conflicted feelings for Catalina, a reluctant celebrity. Like the roosting sites, she didn't ask to be found, and yet in monarch butterfly circles she was a subject of interest. A celebrity. In our conversations over the years, she vacillated between open to guarded and at one point confided that one of the main reasons she withdrew from the monarch story was that she felt completely disregarded in its initial telling.

After we published her story, journalists came calling, but she seemed largely disinterested. In September 2012 when National Public Radio interviewed her, I sent her a link to the audio, which she hadn't heard. After *Flight of the Butterflies* was released and gained traction, she emerged as an unsung heroine of the discovery story. Upon googling her name, producers, authors, podcasters, and journalists found her coming-out story on the Texas Butterfly Ranch website and reached out to me. That continues today.

These content producers always ask if I can "connect" them with Catalina. A producer for American Public Media, a children's book author, a new media celebrity with a science podcast, a Mexican writer editing a collection of essays about heroic women—each would see the story on the Texas Butterfly Ranch website and ask to be put in touch with Catalina. I would never provide her contact information without first asking her permission.

Many of these writers and journalists who called, emailed, or reached out via social media reminded me of what Bob and I classified as *paracaidistas*, or parachuters—journalists we encountered during our time in Central America. After we first married, we spent four years in Costa Rica and El Salvador. Bob worked as a war correspondent for the *Dallas Times Herald* where we had met, and later for *Newsweek* magazine. I freelanced feature stories.

At the time, most papers and news organizations covered the Central American strife from Mexico City. But as the region heated up, a few newspapers committed to full-time correspondents in San Salvador.

During those early years, many journalists blew through the region. Reporters from all the TV networks in the United States—CBS, NBC, ABC, PBS—dropped into town. Most didn't speak Spanish. They arrived wearing khaki outfits with lots of pockets and zippers, equipped with cash, translators, fixers, drivers. They fertilized an entire economy on the second floor and the bar of the Hotel Camino Real in San Salvador. The few of us who lived there made friends with select locals and observed the visiting journalists with either respect or disdain, depending on how they behaved. Many did their best reporting from the bar.

The paracaidistas would touch down at the airport having done little or no homework on their assignment and gravitate immediately to the Camino Real. The bar was the place to absorb gossip or hear what other reporters were doing. Where did everyone go today? Did you see some bang-bang (the common parlance for witnessing gunfire)? Who'd you talk to? Can you share sources?

Bob and I met and befriended every major correspondent during the 1980s who came to El Salvador. Bob shared more than a few dangerous experiences in the mountains with several of them who lived in and worked the region: Chris Dickey with the *Washington Post*, James LeMoyne with the *New York Times*, Clifford Krause with the *Christian Science Monitor*, Julia Preston and Sam Dillon with the *Miami Herald*. These people lived and worked with us. I can't remember the names of the paracaidistas. But I remember what it felt like to be used by them.

That feeling came back to me whenever Catalina hesitated to share specifics of her version of events and engage with the media. After her story was first published, the movie came out, and the monarch butterfly went mainstream, she felt that everyone was benefiting financially from the monarch butterfly story except her. The requests from a new set of paracaidistas began—"Hello, I'm writing a children's book on the monarch butterfly and would love to get in touch with Catalina Trail. Can you help?"

In various conversations over the years, Catalina expressed disappointment at how the discovery story continues to be told. She said that even since the release of the movie the story had been unfairly and inaccurately portrayed. "We're getting closer," she said. "The current version is more accurate, but so many decades have a way of eroding the truth."

One example of Catalina's dismay resulted from how Chip Taylor of Monarch Watch had presented the story of the historic day in 1975 when she and Kenneth had chanced upon the roosting sites. The story was published in 1999, but more than a decade and a half later it still burned Catalina. In it, Taylor described an outing he had undertaken with a Japanese film crew in the fall of 1998 to the roosting sites. Titled "In Pursuit of a Little History," the first-person narrative appeared in the May 1999 edition of the Monarch Watch season summary newsletter.[5]

Taylor shared that he and a film crew from NHK, Japan's equivalent of public television, had filmed an episode on the monarch migration, a project of the network's *Our Living Earth* nature series. They had started in Kansas, traveled to Texas, visited the roosting sites, then returned to the Lone Star State to interview Ken Brugger in Austin. In the account, Taylor described how Brugger had provided him with slides of the historic day. He also wrote that upon visiting Macheros, a small village near Cerro Pelón where Catalina and Ken had finally found the monarchs, he had tracked down the elder campesino who, with his horse, had accompanied Catalina and Ken up the mountain on January 2, 1975.

Taylor wrote in the article that the almost centenarian was named Benito Juárez. "We hadn't expected to find Ken Brugger's guide since we had been told that all the original participants were now deceased. . . . With great anticipation, we marched down the hill in search of Don Benito Juárez. We found Don Benito, an active 96 year-old, at his home. He graciously gave us a long, on-camera interview."

Catalina was offended by this account on multiple levels. She was not deceased. She was there and had actually led the expedition. The local who provided the horse was named Agapito, not Benito Juárez, the first Indigenous president of Mexico, elected in 1872. Agapito was not a "guide" but someone who accompanied them up the mountain. And the "borrowed" slides referenced by Taylor had belonged not to Ken Brugger but to Catalina. As best she could tell, they had never been returned.

"'I do not know if you have ever been dismissed like I was back then, but I hope you understand my perspective," she said, describing her own feelings upon reading Taylor's story.

With the recent recasting of the narrative noting her significant role, Catalina decided to call Taylor out and demand corrections of the 1999 article. In 2016 she began an email exchange with him, insisting he retract the story, correct the record, and return her property, more than two hundred slides.

After some back and forth, Taylor responded via email in July 2016. He committed to retracting the 1998 article and issuing an apology. He said Monarch Watch would provide links to current articles and interviews that reflected her role in the historic day more accurately. And he promised to send a flash drive to her with the more than two hundred slides "loaned" to him by Ken Brugger.

By way of explanation and understanding, Taylor stated in other emails that he did not set up or seek out the interviews with Ken and the so-called guide in Macheros. The NHK producers had arranged that. He also noted that he had no idea how to find Catalina since she had been "under the radar" (her words) for decades. And he claimed he was unaware the slides belonged to her.

Taylor admitted to careless writing and that the "discovery story" had always been told from the point of view of Ken and Urquhart. "Your role was not fully recognized in the Nat Geo article and in many other accounts. I certainly wasn't aware of your role in the discovery when I wrote the article in 1998," Taylor wrote. "I'm glad you are now communicating with reporters and others to set the record straight. You should continue to do so." Taylor was appropriately humbled and reflective, admitting, "I should have been more skeptical, but I wasn't. Lesson learned."

In closing, Taylor stated that his information was incorrect: "That's it. There was no intent to diminish you or Ken or the 'Discovery. . . . Had we known more about your role in this discovery, we certainly would have highlighted your contribution in the article."

On December 5, 2016, Taylor published "In Pursuit of a Little History: A Retraction" on the Monarch Watch blog.[6] The 783-word article began with him owning his mistake and followed with an apology. "Mea culpa is a Latin phrase that means 'through my fault' and is an acknowledgement of having done wrong," he started, finishing the lead paragraph with "I also wish to apologize to Cathy Brugger, as she was then known, and now as Catalina Trail, for the publication of this article."

The article offered a reckoning of the facts as we know them. As promised, he also provided links to interviews and stories that accurately reflect Catalina's role in enlightening the world about the monarchs' winter roosts.

▼

Over the years I have known her, Catalina has mentioned to me the importance of truth and her desire to write her own story. On multiple occasions she has mentioned a desire to write a memoir from the handwritten notebooks she kept from her days traipsing through the Mexican mountains in search of monarch butterflies. Her expressed desire to write an autobiography led me to send her a copy of *The Art of Memoir*, a master class on the elements of great memoir by bestselling author and renowned professor Mary Karr.

I can't wait to read Catalina's book, if she ever writes it. She told me she wants the world to know that in the early 1970s in rural Mexico a young woman was looking to the sky, doing fantastic things and making monarch butterfly history.

▼ ▼ ▼

CHAPTER 11

BETTER THAN CHURCH

My ninety-three-year-old father John Maeckle, affectionately known as Opa, passed away August 2, 2015, after a long battle with dementia. He and my mother Hilde had lived behind us in the casita for a little more than a year before our family decided to move him to a senior care facility. When Opa started soiling himself daily and my mother's noble attempts to keep up with his care made getting a decent night's sleep impossible, we knew it was time. Reluctantly, we delivered my father to Chandler House, a nearby senior care community that offered independent and assisted living as well as acute care.

My parents' casita anchored the south side of my most recent turf-to-bed conversion—the double lot of a new home designed by our son in downtown San Antonio. It sat across from the historic Arsenal, which was built in 1859, the complex of limestone government buildings that served as an arms and munitions depot for Texas frontier battles as well as the American Civil War and both world wars. This latest pollinator habitat evolved across the street from the historic structures, under the shade of a dozen mature pecan trees, volunteers from the San Antonio River bottom just a block and a half away.

Just as we began installing a pollinator garden at our home, we moved Opa to the assisted living facility. Planning and planting the garden became a refuge for me. In between work, settling in, and installing the garden, I did my best to visit Opa at Chandler every day. The facility was conveniently located, just a short drive from the house, and the staff were fantastic and caring. Still, I found visiting my father there extremely depressing. Each trek from the parking lot through the glass doors of the former mansion required deep breaths and a self-pep talk. The stride down the hallways to Opa's shared second-floor bedroom in Room 328 became an exercise in self-composure, as

I passed the aging elders sitting outside their rooms in wheelchairs wearing sad looks of despair and clinging to life via respirators. Daughters and sons, husbands and wives, family members and friends, nurses and therapists— many caring people roamed the halls and provided love and care to these old souls, whose will to live resisted the sad fact that all good lives come to an end.

Opa lasted only four months at Chandler. In his final days, the spirit of the man who taught me to find solace and adventure in nature and that "life is full of compromises" had fled. His body appeared a shell of the vibrant outdoorsman, as empty and lifeless as the chrysalis casing left behind after a monarch takes first flight.

Just days before he passed, I visited the old man and whispered in his ear, "It's okay to let go." He was no quitter, and I don't recall him ever asking permission from anyone to do anything. But somehow, I think he considered my suggestion. Not long thereafter, he died. He took off August 2, 2015, hours after we had all paid him a visit.

We politely declined the offer by Chandler's resident pastor to say last rites for Opa, who preferred Sundays at the lake or deer lease over church. His body was retrieved by ambulance, taken to the mortuary, and prepared for cremation.

As first-generation Americans, our family had no firsthand experience with death. No one in our immediate family had ever died. My mother learned of my grandmother's passing via *Luftpost*, German airmail, just two years after she arrived in America in November 1953 from the northern German seaport of Bremerhaven on a transatlantic freighter called the *Jesse Lykes*. A long-distance phone call advised my father that his mother had passed away on the same Schwabian farm he had left as a young man. The news of other faraway kin dying usually was announced conversationally. "Uncle Dan passed away," or "Ruth died," my mother relayed plainly by phone or when our paths crossed in person. "Oh. Sorry to hear that."

This inexperience with death posed a question for our family: how should we celebrate Opa's life? Given our shared disdain for organized religion and my passion for butterflies, we decided a butterfly release would be appropriate. For us, a church service or funeral home reception would feel like posturing. Decades of living in San Antonio had also exposed us to Día de los Muertos celebrations, whereby death is honored as a natural stage in the life cycle.

▼

I had learned a lot about butterfly releases through my brief detour into commercial butterfly breeding. Although my career lasted only a few months and included my single escapade at the Insecta Fiesta, I had studied the business for three years, served on the board of the IBBA, attended their conferences, and taken best practices classes on rearing healthy, disease-free butterflies. I had also read thousands of emails addressing the challenges of running a successful butterfly breeding business. And yes, during that education I had learned how extremely controversial commercial butterfly breeding and releases are for certain segments of the scientific and enthusiast community.

My IBBA friends were understandably tired of the anti-breeder arguments, which appear to have entered the mainstream via the *New York Times* in 1998. The article, "Festive Release of Butterflies Puts Trouble in the Air," provoked a national discussion of the appropriateness—or not—of commercially bred butterflies being released into the universe.[1] In the late 1990s, Rick Mikula's novel idea of butterfly releases at weddings, funerals, and celebratory events was gaining popularity. The article raised the flag on the notion that using butterflies in such a way could be misguided and wrong.

Jeffrey Glassberg, founder of the National Butterfly Center and proponent of appreciating butterflies through binoculars, called the practice "absolutely disgusting" and the "worst thing you could do at your wedding." More than twenty years later, Glassberg, who travels the world leading butterfly tours, still views commercial butterfly breeding and mass releases as misguided arrogance, "a travesty."[2] In particular, he believes that the U.S. Fish and Wildlife Service wastes millions and millions of dollars paying zoos to raise butterflies that they then "dump back into the environment. . . . And then they all die, and nothing ever changes . . . because they never spent any time and effort figuring out why the butterfly declined in the first place."

In the *New York Times* article, still cited two decades after its publication, Robert Michael Pyle, who spoke at the IBBA meeting in 2006, accused commercial butterfly breeders of "turning butterflies into baubles." The resulting public debate among butterfly lovers continues to this day. In varying degrees, the National Wildlife Federation, Monarch Joint Venture, the Xerces Society, and, yes, PETA all join Glassberg and Pyle in condemning the release of commercially bred butterflies into the wild.

One of the main concerns, as noted earlier, is the spread of disease from commercially bred butterflies to those in the wild. Critics contend that the high density of butterflies sometimes found in captive breeding situations at commercial butterfly farms increases diseases. In particular, the anti-commercial breeding contingency believes that OE, the vexing monarch-centric spore-driven disease vector *Ophryocystis elektroscirrha*, would more likely be found in commercial breeding operations because of the high concentration of butterflies in confined spaces. When these disease-carrying monarchs enter the wild population after being released at a wedding or a funeral, they could drop their spores on a milkweed plant that might later be consumed by a wild monarch caterpillar. The entire migrating population could be placed at greater risk.

Some scientists also contended that commercially reared butterflies could corrupt their annual butterfly counts and migration tracking and disease studies, which could interfere with conservation efforts. Proper scientific tallies would become inconvenient if not impossible. "It's unnecessarily muddling the biology of the monarch butterfly," said Lincoln Brower.

In one testy email exchange in 1997, Pyle chastised butterfly zealot Hans Schnauber, founder of the now defunct International Butterfly Federation of Enthusiasts and promoter of a nationwide July 4th National Butterfly Release Day decades ago. Schnauber had released 495 reared monarchs upstream from Pyle's usual collection destination, and Pyle was not happy about it. He accused Schnauber of polluting the skies with home-bred butterflies along the California coast and compared the act to "rascalism," a term he said butterfly farmers in Papua New Guinea used for troublemakers. "I think that applies nicely here. With so many serious questions of natural history and conservation challenges to occupy our time, energy and enthusiasm, I hope other monarch lovers will join me in repudiating such activity," Pyle wrote on the D-PLEX on September 29, 1997.

But passionate and informed voices rose on the other side, too. Edith Smith, one of the largest commercial breeders in the country, pointed out that, in the more than two decades since commercial butterfly breeding and releases appeared on the radar of those who oppose it, not a single case of disease spreading from a released, commercially bred butterfly to wild stock has been documented. Smith, her husband Stephen, and several family members owned and operated Shady Oak Butterfly Farm in Florida from 1999 to 2022.[3] For more than two decades, the enterprise raised and shipped mega-thousands of butterflies around the country.

Those who suggest the introduction of disease as a concern counter that the absence of data or a specific example doesn't mean the release of commercially reared butterflies has zero effect on the wild population. It just means that it has not been studied and may, in fact, be too difficult to study.

Smith takes special offense at the "butterflies as baubles" characterization. "They are not treated as baubles, not by a long shot," she says with measured emotion. "Baubles are never 'made' nor treated with such love and concern." She couches her butterfly zeal in soft-spoken, well-informed experience and often praises the Lord for a life she admits has enjoyed many blessings. A self-taught expert on butterfly breeding, she is well known for the generosity of spirit with which she shares her butterfly knowledge. Her website, Butterfly-Fun-Facts, is packed with useful how-to information on rearing plants and butterflies, pest management, and only the occasional plug for her breeding operation.

For years she conducted an internship program at her Florida home/farm, which hosted more than a dozen guest interns from fourteen countries who hoped to learn best practices from a veteran breeder. Smith also collaborated with pathologists, lepidopterists, government officials, and conservationists to develop and teach an online course conducted by email and housed online at Butterfly College, a project of the Association for Butterflies, a butterfly conservation and education organization that split off from the IBBA in 2004. The college offers free classes to funeral professionals and wedding planners about how to conduct butterfly releases. For members, the college offers more in-depth courses for butterfly farmers on how to pack and ship butterflies as well as how to recognize and prevent disease when breeding butterflies.

Smith wonders why there's such an uproar about butterflies being shipped across state lines and unleashed into nature when the nursery industry moves tens of thousands of pots of milkweed and other host plants from state to state each season unchecked. She contends that the potential for spreading disease through transported plants is exponentially greater than disease purveyance by butterfly farmers. Many plants are shipped from the warm American South and California to all parts of the country. The OE found in western monarchs is more virulent than the OE found in the eastern population, Smith points out. In addition, the prevalence of OE is highest in warm, southern states, especially in the Miami/Dade area, which is full of nurseries that ship milkweed north.

Passionflower vine, host plant to the Gulf fritillary, is also shipped all over the country. Passionflower often includes cells of the Nuclear polyhedrosis virus (NPV), says Smith. The cells are left by dead Gulf fritillary caterpillars that feast on its leaves. NPV is 100 percent fatal to caterpillars, and unlike OE it is not specific to a particular species of butterfly. It can be picked up from any infected caterpillar's regurgitation or excrement. Like the much-studied OE spore affliction, NPV is unleashed in the stomach of the caterpillar, where the acidic environment activates it. It attacks the larva's cell structure. Crystal-like formations grow, enlarge, and fill up the creature's cells, causing them to explode. After a bout of sluggishness, diarrhea, and the vomiting of fluids, the caterpillar eventually "melts" like the wicked witch of the west. The contaminated body fluids that contain the virus can then be picked up from plant leaves by other caterpillars or butterflies, which then spread them to other plants.

And yet . . . anyone can go online for a late afternoon plant shopping spree and order an exotic, organic "New and Healthy Yellow Passion Fruit, *Passiflora edulis*, Potted Starter plant Tropical Vine" from Jupiter, Florida. Add it to your cart and type in your credit card info and it will be shipped cross country from Florida to Puget Sound, if you like. No questions asked. NPV cells, if present, no extra charge.

"It's like being worried about pneumonia when Ebola is around," says Smith, who's trained as a horticulturist. Well-versed in plant physiology and butterfly science, Smith likes to experiment at her farm. Through selective breeding, she developed the blue buckeye, *Junonia coenia*, a version of the classic buckeye butterfly that flaunts showy blue and green scales. The only manipulation used to create these beauties was simply choosing which females to mate. She and her daughter also cultivated a new strain of variegated milkweed, 'Charlotte's Blush.'

Mild-mannered Smith, who professes a belief in respectful disagreement, thinks the nursery business is too well funded for butterfly release opponents to attack. "In their opinion, we show disrespect for butterflies while they don't believe nurseries do that," she says. "Shipping plants is okay to them, but not butterflies. If disease was the real concern, they'd be after nurseries, too."

Counterarguments to those against commercial butterfly breeding and releases also include the fact that wild salmon and trout seem to do just fine when the coddled commercially bred are introduced into streams. In fact, adding new, healthy DNA to the gene pool may even strengthen populations in decline.

In October 2017 at our second annual Monarch Butterfly and Pollinator Festival, we released 720 monarch butterflies commercially bred in Bradenton, Florida, by Connie Hodsdon of Flutterby Gardens. The butterflies were tagged and their data recorded and sent to Monarch Watch. Five months later, we learned that five of our charges were recovered on the forest floor of El Rosario Sanctuary in Michoacán. The incident provoked hope, amazement, and a certain testament for many monarch enthusiasts who have been told by some monarch scientists that commercially reared monarchs don't have the Darwinian chops to make it to Mexico.

"And now evidence shows that commercially reared monarchs can, and some do, successfully migrate to the Mexican overwintering grounds," said monarch butterfly researcher and chemical ecologist at Cornell University Anurag Agrawal, upon learning of the news. Also hearing the news just months before his death in 2018, Lincoln Brower, a staunch critic of commercial breeding and mass releases, said that the recovery of reared monarchs "speaks for itself." "One might quibble over the conditions they were reared under, but at least under certain conditions they can be recaptured over long distance."

Since 2017, five more butterflies released at our festival have been recovered in Mexico. Examples of commercially bred monarchs released at other public events have also occurred, begging the question, could commercial breeding play a role in restoring the monarch butterfly population?

Widely publicized incidents of dead or dormant butterflies released at public events have not helped breeders' gain acceptance by some in the conservation community. Certainly bad breeders exist, but user error is often the culprit behind butterfly release fiascos. Breeders include very specific instructions when shipping butterflies—keep the ice packs fresh, don't leave the box in full sun, give the butterflies time to warm up, and feed them if necessary.

One famous example that lit up social media: a disastrous stunt attempted by performer Asia O'Hara, who was competing in RuPaul's Drag Race, a reality TV show that documented former supermodel and America's most famous drag queen RuPaul conducting a search for "America's next drag superstar." O'Hara performed a dance wearing a "butterfly bra" in which painted lady butterflies were stuffed. The small net chest cones, poised erectly on the chest of the dancer, each held dozens of butterflies. While dancing to Janet Jackson's "Nasty," O'Hara attempted to unleash the leps, but . . . nothing happened. As Yahoo Entertainment news reported, "Asia tore off one butterfly-boob and then the other, tossing both desperately into the air, only to reveal more dead or dying bugs, many of which she then stomped on while working the stage."[4] The video made the rounds as an "epic butterfly fail" and "butterfly fiasco," and PETA later described the incident as a "sad spectacle."

Less flashy disasters include birds picking off painted lady butterflies released at an elementary schoolyard, monarch butterflies left in the blazing sun before a funeral service at which they were supposed to be released but instead expired from the heat, and not quite warmed-up butterflies dropping to the ground and getting crushed by spectators. These unfortunate events don't help breeders' case, even though those who bred and delivered the butterflies were not to blame for these fiascos.

Of equal and perhaps greater concern are the legions of amateurs who raise hundreds, sometimes thousands of monarch butterflies each season, unleashing them into the universe. These members of the Monarchy, often newcomers to monarch butterflies, become captivated by the miracle of metamorphosis they can witness by home rearing. They bring eggs or caterpillars in from the wild for fostering. When health problems arise with their butterfly charges, these home breeders reach out to experts and social media for answers. Unpleasant advice—such as the need to euthanize sick butterflies rather than release them into the gene pool—is not well received.

The conflict came into focus in the fall of 2018 when the Xerces Society published a blog post, "Keep Monarchs Wild! Why Captive Rearing Isn't the Way to Help Monarchs."[5] Xerces is known for its opposition to commercial butterfly breeding and butterfly releases as well as its objections to tropical milkweed as a host plant. The organization also served as one of several authors on the 2014 petition to list the monarch butterfly as threatened under the Endangered Species Act (see chapter 7).

While recognizing the power of rearing butterflies at home to engage the public, Xerces's "Keep Monarchs Wild!" article raised legitimate concerns about captive rearing and the dangers that occur when people capture wild butterflies and breed them repeatedly, the biological equivalent of inbreeding. Xerces conservation biologist Emma Pelton shared a link to the post with the subject line "Rethinking captive breeding" on the D-PLEX list, and the debate unfurled.

Captive rearing of monarch butterflies can have unintended consequences, Pelton pointed out, including spreading parasites to wild populations and diminishing genetic diversity. She suggested that, instead of rearing butterflies, butterfly lovers spend their energies improving habitat, avoiding pesticides, planting native milkweed and flowers, and supporting wildlife-friendly agriculture and policy initiatives.

Within a week, dozens of comments and counterarguments had been posted. The online kerfuffle included takes from monarch scientists. Several D-PLEX subscribers asked to be removed from the list, declaring that their forum had been "hijacked." "Our position at Monarch Watch is that we neither encourage nor discourage rearing," wrote Chip Taylor. Taylor's citizen science organization has a page on its website devoted to raising monarchs and also offers Monarch Rearing Kits for sale. "This is a low priority issue," he said.

Monarch Joint Venture labeled the practice of raising small numbers of monarchs for citizen science, educational purposes, or personal enjoyment "great applications, as long as done responsibly!" The Xerces post included a link to a recently updated Monarch Joint Venture handout that provided specific guidelines for responsible home rearing. The update came in response to the hundreds of calls and emails Monarch Joint Venture was receiving each month asking for direction about home rearing, outreach coordinator Wendy Caldwell wrote in a statement.

In a follow-up email to D-PLEX subscribers, Xerces's Pelton provided context for the post: "I continue to end up with an in-box full of horror stories: people driving monarchs to overwintering sites because they raised them too late in the year, moving monarchs across huge geographic areas, caterpillars starving when there's not enough milkweed in urban areas paired with massive captive-rearing operations, hundreds of monarchs raised en masse, photos of diseased & sick monarchs, etc."

I can attest to Pelton's woes. I often receive emails and social media messages asking me if I'd be willing to accept butterflies raised in other parts of the country—typically places farther north where the weather has turned cold. Typical messages: "Hello. My name is Victoria. . . . My neighbor has monarch chrysalises that are going to enclose [sic] soon. We are in Chicago and it may be too cold to release them here. Would you be interested in taking them to release further south?" and "I really hate to ask this, but do you know anyone visiting the D/FW area this weekend? Unfortunately, I have Monarch caterpillars because I didn't cut back my milkweed in November (long story, but don't get mad at me) and I would like for someone to give them a ride South since it's supposed to get really cold next week."

This overzealousness is not new. In 2011, "Butterfly Lady" Maraleen Manos-Jones of Shokan, New York, author of *The Spirit of Butterflies*, a book that explores the otherworldly dimensions of Lepidoptera, convinced Southwest Airlines to comp a 1,950-mile flight from her chilly home up north to San Antonio. Manos-Jones had a monarch that eclosed late in the season with temperatures in the 40s. The goal of her comped flight? To allow the creature to play "catch up" on its southbound migration.[6]

Other scientists have bluntly expressed the sentiment that raising butterflies at home doesn't help the migration. Andy Davis, an expert on migration, posted an article on his blog with the headline "New Statement from Monarch Conservation Groups Says—For the Love of God, Stop Mass-Rearing." Davis cited various scientific publications that suggest home-reared monarchs are less healthy and less likely to make it to Mexico.[7]

Then the author of one of the cited articles, Gayle Steffy, chimed in that her research was being taken out of context. "Since my paper has been quoted and debated here, I'm adding my opinion, which is that the data do not indicate that wild monarchs are better than raised ones," Steffy said.

At one point, ecologist Agrawal declared soberly: "Let's not kid ourselves that we are helping the population by rearing. The sink full of water has a leak, and yes, we can add a few drops back, but only plugging the leak will help.

Sorry for the pessimism, but mass rearing is a flawed strategy for monarch conservation."

For better and for worse, those seeking to learn about raising monarchs responsibly are often unaware of or overlooking scientific websites or the D-PLEX list in favor of social media destinations like The Beautiful Monarch, a public Facebook group with more than 100,000 members.[8] It's too easy to drop into these sites, ask a question, or post a photo and come back later for an answer without investing too much time or energy.

David Berman, an Oklahoma State University biologist, focused on parasitism in monarchs, waded into the social media quagmire on the Beautiful Monarch Facebook page one day back in 2018. Berman was hired to conduct research on tachinid flies and milkweed availability in Texas. Periodically, he would swing through the Texas Hill Country to monitor various milkweed patches, including the swamp milkweed that grows along our stretch of the Llano River. Beautiful Monarch group member Georgi Grey of central Ohio posted a photo of a slide viewed from a microscope, showing butterfly scales infected with OE spores.

People who raise monarch butterflies at home are encouraged to test for OE by rolling the butterfly's abdomen on a piece of tape to get a spore count, then looking at the tape under a microscope to see if the butterfly is infected before releasing it to the wild. "How bad is too bad to release? Here are the photos I took." Grey posted in a comment to the group that accompanied her photo: "I just started testing for OE."

Berman happened to be checking Facebook at the time and responded that releasing butterflies with OE is *not* a good idea. Group member Cari Waller then countered, "If the butterfly is still strong, no falling, no trouble flying, no split proboscis, etc., then it is ok to release them." A split proboscis, by the way, is another butterfly malady. The proboscis, the "strawlike" body part butterflies use to consume nectar, forms in two interlocking pieces, which should zip into one after hatching to form a functional feeding tube. Sometimes this doesn't happen, resulting in a nonfunctional "split proboscis."

The online conversation spiraled into dozens of comments of conflicting advice. Some group members insisted that, as long as milkweed is bleached and the OE spore load "reasonable," it was safe to release the OE-infected

butterflies into the wild. Others assumed incorrectly that OE can be transferred only from the plant or a parent. Within twenty-four hours, admins of
the group cut off the conversation because "it got too heated," said Berman,
who continued to receive private messages about the subject.

He patiently explained that spores are sticky and adhere to plants. They
"can't be easily washed off. That's why bleach is recommended." They also
cling to your hands, he said, adding that a colleague found spores on his
hands more than a week after handling OE-infected butterflies in the lab. "I
have accidentally contaminated lab butterflies more times than I'm proud
to admit," he wrote.

Ultimately, Berman appeared to influence at least some in the crowd,
including Grey. "I'm not sure I totally understand OE at this point," Grey
said via social media:

> I've read conflicting stuff. I euthanized two that were heavily infected, yet
> flew fine, after that discussion—but that was before I'd heard from Project
> Monarch Health saying they didn't advise euthanizing. However, on their
> own website I found info to the contrary! So now, I don't really know what
> to do. I tested a few more, found two more with OE, but not as heavily
> infected, so I went ahead and released them. And have worried about that
> decision! And then I stopped testing, partly because I just didn't have time.

In this messy context, commercial breeders with protocols for sanitation
and standard operating procedures sound like more of a solution than a
threat. They make a living breeding butterflies, and to do so, to stay in business, they must maintain a clean shop.

Shoddy breeders exist, no doubt. But ignorant customers that don't bother
to read the handling instructions and well-intentioned amateur breeders
who fail to learn basic biology and best practices should also be considered
villains for mishandling precious butterfly livestock. Reputable breeders are
quite clear: don't leave butterflies in the hot sun to cook before the release.
Treat them with the respect they deserve.

▼

In the case of Opa's life celebration, I considered these conflicting viewpoints
before deciding with my family that a butterfly release would be the ideal

Ninety-three butterflies for Opa's life celebration.
Photo by Monika Maeckle.

gesture for commemorating my father's life. I ordered ninety-three butterflies from Barbara Dorf and Tracy Villareal of Big Tree Butterflies in Rockport, Texas.[9] I knew it was preferable to stay within your geographic area, if possible, when ordering butterflies. It's best for the butterflies and limits their travel time. The cues from the sun remain when they are reared and released in the same general latitude. I had come to know the couple through my time at the IBBA. At the time, Barbara and Tracy bred butterflies on the side while also working as professional scientists at the University of Texas at Austin. They followed the best practices I had learned while exploring the possibility of becoming a professional breeder.

The butterflies arrived early that sunny August Friday, five days after Opa seized his last breath. Typically, breeders pack butterflies in the afternoon for overnight shipment and early arrival the following day. Opa's butterflies arrived by 10:30 A.M. from the Texas Gulf Coast, about a two-and-a-half-hour drive south of San Antonio. Each butterfly rested in its own glassine envelope, set inside a cardboard box, carefully packed with blue ice packs wrapped in paper towels. The glassine envelopes, made from a mix of tissue and wax paper, allow for the passage of air. The ice keeps the butterflies cool and dormant, and the paper towels absorb any condensation from the ice packs.

After unpacking the butterflies, I placed about thirty of them in a pop-up cage (actually a converted laundry hamper), spritzed the white, zippered, net-covered container with water, and set them on the dining room table where the temperature was a comfy 70 degrees. The rest of the butterflies remained safely in the air-conditioned darkness of my closet, waiting for the appropriate moment to take to the skies.

Friends and family started arriving right before 5 p.m. My brother Mike Maeckle had flown in from Montana and would stay with us for a few days. My mom walked the thirty-three steps from the casita to join us in the kitchen as people trickled in.

The "service" was straightforward and brief. After a poem by our neighbor poet Naomi Nye, my niece Melinda Maeckle Martin and her husband Robert, both trained as opera singers, performed a duet in German. Family members recalled fond memories of Opa. And then we asked everyone to move from the cool of the living room into the late summer sun of the butterfly garden. Each of about fifty friends and family received a healthy monarch butterfly in its protective envelope—the same type of enclosure Lincoln Brower used to store the butterflies he collected at the ranch.

After a beautiful baritone rendition of the German version of "Taps" sung by Robert Martin, we counted to three, in German, "*Eins, zwei, drei!*" Each person opened the loose flap of their envelope to release their butterfly. Off they went. We unzipped the net cage, and the remaining butterflies joined the flight.

Ooos and ahs filled the yard as guests of all ages, from toddlers to seniors, marveled. The butterflies lilted on shoulders, danced on the daisies and milkweeds, and drifted around the yard in their dreamy flight pattern like so many old souls. It was easy to understand why the Indigenous peoples of Mexico think of monarchs as their ancestors returning for a visit each fall when they arrive in early November for Día de los Muertos. The belief that death is a part of the life cycle to be accepted and celebrated was also somewhat reassuring. My mom, eighty-two years old at the time, described the celebration as better than any church service.

Because my butterfly garden provided plenty of nectar sources as well as five different types of milkweeds that summer, the monarchs stuck around for a while. A week later, half a dozen were still enjoying the yard, some of them laying eggs, which later hatched. We tagged them and sent them south. I so wished for one of Opa's monarchs to be recovered on the forest floor in Mexico. So far, that hasn't happened. I'm still hoping.

This tactile experience caused dozens of friends and family who had never thought twice about butterflies to now view them differently. The children in attendance chased butterflies for hours after the release, completely enchanted, forever touched by them. The joyous faces looking skyward shown in photos taken moments after the release convey the reason many butterfly breeders continue practicing their profession: it's gratifying to spread joy. Breeder Edith Smith tells a story of a funeral director who participated in a butterfly release for the first time. At first he was skeptical. But after the ceremony he relayed that it was the first funeral he had ever attended at which each participant walked away from the casket wearing a smile.

Conservationists will tell that you the most effective path to protecting a species or an ecosystem is engagement. Paying attention is the first step to understanding. Understanding begets caring and that evolves into action. Butterflies are masters of engagement. And that engagement and the consequent embrace of their conservation outweigh the unproven risks claimed by butterfly release detractors.

As for speculative claims that thousands of commercially raised butterflies released into the ecosystem will "pollute" the gene pool, there's no evidence of that to date. And the numbers just don't add up. One IBBA member shared an internal survey of breeders done in 2015 that suggested fewer than 250,000 monarch butterflies are released each year around the United States—about 0.28 percent of the approximately 85 million monarch butterflies estimated to have migrated that season.[10] Those butterflies are also released throughout the year—at different times and in different locations throughout the 3.8 million square miles of the United States. These numbers don't take into account the thousands and thousands of butterflies that amateur home breeders are contributing to the skies. Chip Taylor and other scientists have suggested repeatedly that, given current numbers, a focus on butterfly releases is a misplaced priority.

Do the joy and engagement with butterflies and the resulting embrace of their conservation at a time when people are increasingly removed from nature outweigh the unproven risks claimed by butterfly release detractors? The debate and science will continue to try to answer that question.

But I know what Opa would say. "Life is full of compromises. You just have to decide which ones you're willing to make."

▼ ▼ ▼

CHAPTER 12

THE MILKWEED QUANDARY

A pair of monarch butterflies sailed across the San Antonio River just a short walk from the Pearl Brewery on a warm afternoon in January 2012. The orange-and-black visitors paused on tropical milkweed flowers, the star attraction of San Antonio's "Milkweed Patch" on the city's lauded Museum Reach riparian restoration. Locals of all ages wandered the paved trail, one of many along the thirteen-mile, $384 million San Antonio River Improvements project begun in 1998 and completed in 2013. The project earned the prestigious Thiess International Riverprize in 2017 and no doubt contributed significantly to the unanimous 2015 vote in Bonn, Germany, that declared San Antonio's Alamo and Spanish colonial missions built along the river a UNESCO World Heritage site, the first in Texas.

A natural gas-powered tourist barge navigated the quiet waters as one of the monarchs, a female, landed on a lone orange milkweed flower. Poised on the showy bloom, she tucked her abdomen, as if doing standing sit-ups, to reach the leaf's underside. Carefully, she oviposited a single creamy white egg. In a few days, a tiny caterpillar would emerge and start consuming milkweed leaves. It would morph through its stages to realize its signature bright green chrysalis. Barring a freeze, predators, or illness, a butterfly would emerge within the month. And in this especially warm, dry winter—at the time, the hottest and driest in recorded history—dozens of monarch caterpillars in various stages noshed on milkweeds at what had become an urban butterfly nursery. Spent chrysalis shells dangled from the concrete overhang of the Milkweed Patch's deck, like forgotten Christmas ornaments.

The Milkweed Patch inspired visitors and heartened locals. Not long ago, trash, vagrants, and invasive weeds dominated this long neglected riverscape. Now, resident caterpillars and butterflies—monarchs and others—occupied its refurbished banks, a living metaphor of transformation and possibility.

And yet the milkweed species at the Milkweed Patch, *Asclepias curassavica*, commonly known as tropical milkweed, provokes vigorous national debate. In monarch butterfly circles, the plant's growing popularity and visibility beg questions of what is "native," underscore our human influence on nature and the monarch migration, and ask how far we should go to ensure the continuation of a unique and arduous migration that by many measures may have already outlived its usefulness to the species that undertakes it.

As its name suggests, tropical milkweed comes from the tropics and is technically not native to the United States. Scientists aren't sure exactly when the plant arrived here from the Caribbean, Mexico, or perhaps South America. Its Latin epithet, *curassavica*, suggests Curaçao as its provenance.[1]

Butterfly student and faithful observer John Abbot documented monarch butterflies noshing on *A. curassavica* in the middle Atlantic United States in a natural history collection published in London in 1797.[2] A survey by Abbot included fifty of the more unusual Lepidoptera species observed in Georgia, as well as their preferred host plants. Beautiful color plates illustrate the leps in their caterpillar, chrysalis, and adult stages, including two views of monarch butterflies—one from above with its uniform pumpkin-and-black colors, the other from below with its tawny gold underwings. A fifth instar monarch caterpillar sits on a thin, half-eaten tropical milkweed leaf, one node up from a chrysalis.

"This caterpillar eats the butterfly weed, *Asclepias curassavica*," wrote the author, using the generic moniker for any plant that attracts butterflies in large numbers. "On the 24th of April it suspended itself by the tail; changed to a chrysalis next day, and on the 11th of May, the butterfly came out. It is not a very common species."

By 1806, botanical journals had suggested that American gardeners were planting tropical milkweed in their yards and greenhouses. A gardening manual published in 1890 touted the glamorous and reliable bloomer as "worthy of a place in our gardens." Milkweed scholar Robert E. Woodson, who assembled a survey of North American species of *Asclepias* back in 1954, noted the presence of milkweeds as "occasional ruderals," or pioneer plants, early to take root in disturbed areas of California, Florida, Louisiana, and Texas.[3]

Commercial growers in Europe were likely experimenting with tropical milkweed throughout the twentieth century. European seed cultivator Kieft Company, based in the Netherlands, introduced the 'Silky Gold' cultivar of *A. curassavica* in the 1990s, a lemony yellow sister of the original orange strain that originated in the tropics. 'Silky Gold' underwent field trials in 1994 en route to its destiny as a winner of the prestigious FleuroStar award in the novelty category. This international recognition is given to the most promising flowering plants each year in a contest curated by professional growers and others associated with the commercial plant growing business.

The same year, the Association of Specialty Cut Flower Growers conducted field trials of the new arrival to test its marketability. Reports from Oklahoma testing sites described the Kieft cultivar "doing very well in warm areas of the country."[4] The yellow-gold blooms showed chill injury when temperatures dropped below 55 degrees. Growers loved its bold, unique colors, long nineteen-inch stems, ease of growth, and status as "something different."

Such plant traits feed favorably into commercial growers' consideration of the "wow factor," every gardening center's spring mandate to stock stunningly beautiful plants that boast the winning combo of bold colors and low maintenance. The FleuroStar contest guidelines spell out the priorities clearly, urging its independent panel of judges—breeders, growers, producers, even journalists—to choose winners with "the 'wow' effect at point of sale." The organization's website continues, "Supported by this prestigious Award, FleuroStar winners become genuine eye-catchers in garden centers and reach millions of consumer gardens."

Depending on where you live, tropical milkweed is considered either an annual or a perennial. In grow zones 8 through 11, the plant blooms and regenerates from spring until winter. Elsewhere it serves as an annual that can be gregarious. After flowering, all *Asclepias* species produce distinctive oblong seed pods. When ripe, the pods bust open to reveal dark gold, oblong seeds, each one attached to a strand of milkweed fluff. The buoyant whitish fluff, which varies in heft and length depending on the milkweed species, can carry the seed elsewhere via wind germination. During World War II, the feather-light fluff was grown commercially for use as stuffing in life vests because of its buoyancy. In short, the ability of milkweed seeds to move, like the monarch butterflies to which the plant plays host, makes the plant a ready traveler.

Tropical milkweed closely resembles the native *Asclepias tuberosa*, often known at gardening centers as butterfly weed. With their similar orange blooms and relatively easy-to-grow habits, the two species are often confused with each other and mislabeled. Side by side, they appear as fraternal twins, yet their chemical composition indicates huge differences.

As discussed earlier, milkweeds contain toxic chemicals—heart-stopping cardiac glycosides, also known as cardenolides. The potency of these chemicals varies greatly, depending on the *Asclepias* species. Tropical milkweed is much more toxic than its doppelganger species butterfly weed despite their similar appearance. In fact, *A. curassavica* is one of the most toxic of all one hundred or so North American milkweed species.

For interesting reasons, monarch butterfly moms, when patrolling the milkweed patch in search of a desirable location to oviposit an egg, almost always choose tropical milkweed when given a choice. Monarch caterpillars also prefer *A. curassavica* as food when presented with a buffet of milkweed options.

Tropical milkweed, the species on which monarch butterflies may have evolved, according to some experts like PhD and cowboy entomologist William Calvert, happens to be absolutely loaded with the pungent chemicals that make monarch butterflies distasteful to predators. Studies show that monarch moms, especially those infected with OE, which they can pass to their offspring when the egg passes through the abdominal cavity to the leaf surface, choose tropical milkweed 68 percent of the time over less poisonous milkweed options.[5] The strong toxic cocktail offered by tropical milkweed inoculates future monarch butterflies to some extent, increasing their resistance to this common monarch-centric parasite. The pungent chemicals proffer a bitter taste to predators, and their absorption into the monarchs' body and expression in bright warning colors provide protection.

One study demonstrated that monarch mothers can lay their eggs on medicinal milkweed that will make their future offspring less sick.[6] Infectious disease biologist and monarch researcher Jaap de Rood characterized the butterflies' choice of tropical milkweed as self-medication.

Monarch butterfly on tropical milkweed,
Asclepias curassavica. Photo by Monika Maeckle

In other words, for monarch butterflies a technically nonnative milkweed creates higher resistance than natives to one of its most serious natural parasites, a parasite that seems to thrive in warmer climates in a world getting hotter. In addition, when consumed, the more toxic milkweed expresses itself more vividly to natural predators, discouraging them from eating the butterflies. Said species is also widely embraced by gardeners because it's affordable, readily available, beautiful, and easy to grow.

Sounds like more of a solution than a problem—yet plenty of people disagree with that conclusion. Ecologist and monarch disease scientist Dara Satterfield, working with Sonia Altizer of the Odum School of Ecology, was among the first to point out that tropical milkweed's persistent growth into the fall and winter allows OE to build up on milkweed plants over the season, creating a hotbed for the very parasite against which the plant creates resistance and tolerance.

For example, a tropical milkweed plant that starts blooming in late March and is decimated down to its stems by hungry caterpillars hatched from first-generation monarch eggs in April will sprout new leaves, stems, and blooms by June. This new growth invites more butterflies to visit. Meanwhile, OE spores, which look like tiny, microscopic footballs, fall like dandruff from the scales of visiting butterflies infected with the parasite onto the milkweed plants. Monarch caterpillars consume the spores when eating the milkweed leaves; adult female monarchs also end up passing the spores to their offspring when the egg passes through their spore-dusted abdomen to be deposited on a milkweed leaf. Thus, the more infected butterflies that visit a particular milkweed plant, the more OE spores build up on the plant.

As the season wears on, spores build up as the resilient *A. curassavica* moves through a cycle of decimation and regeneration. Visiting butterflies drop spores on its leaves; monarch eggs hatch and caterpillars consume the infected fodder. The OE spores lay dormant until a caterpillar consumes and activates the time bomb that OE represents. Since the emerging caterpillar eats its eggshell as a first meal, OE spores left on the egg's creamy white ribs enter its body along with the protein-rich calories of its shell.

Once inside the caterpillar, the spores move through its intestinal tract to the middle part of the gut, where digestive juices dissolve them and release more parasites. These then migrate outside the intestinal wall to the tissue under the caterpillar's skin and divide. By the time the monarch reaches its chrysalis stage, hundreds of daughter parasite cells have reproduced in its DNA. The result: compromised adults emerge weak and deformed from the chrysalis. In severely infected butterflies, contorted and crumpled wings are common. Fatality can result.

Of equal or perhaps even more concern is that the presence of tropical milkweed may encourage monarchs to break their diapause, the temporary asexual state into which the creatures enter each fall. About the time of the fall equinox in mid-September, monarchs get their cues from the sun to head south. They suspend all sexual and reproductive activity and apply their fat stores and energies to migrating. But when tropical milkweed is present along their route, Satterfield and other scientists propose, the monarchs can't resist its hospitality, especially during warm weather. They break their diapause and deposit eggs. Thus, they don't migrate.

Some monarch scientists believe that if the butterflies don't migrate to Mexico the gene pool will be weakened by a lack of diversity. They also worry that OE-infected butterflies that make it to Mexico will contaminate the migratory population and put it at risk. Lincoln Brower famously asserted that tropical milkweed should not be grown outside of a greenhouse any farther south than Orlando, Florida—28.5 degrees latitude.[7] That advice has been widely ignored by the commercial growers and gardening community.

Satterfield visited San Antonio's Milkweed Patch in 2012. An Odum School of Biology student at the time, she collected data on that visit that became part of her 2015 study, which launched a national milkweed ruckus among the Monarchy. Articles in the mainstream and scientific press sensationalized the findings with headlines like "Plan to Save Monarch Butterflies Backfires" and "Are Gardeners' Good Intentions Killing Monarch Butterflies?" Citizen scientists, gardeners, and monarch conservation organizations that had been planting and promoting tropical milkweed wondered if it was truly a bad idea, and, if so, what they were supposed to do since natives were wholly unavailable.

Weighing in with a different view, Jeffrey Glassberg, president and founder of the North American Butterfly Association (NABA) in Mission, Texas, rejects netting and pinning butterflies and advocates that people observe them through close-up binoculars. Glassberg holds a PhD in biology, a law degree, and credentials as an entrepreneur, author, and butterfly advocate. His 1993 book *Butterflies through Binoculars* is credited as the first field guide to encourage pure observation versus netting or pinning of butterflies. His other guidebooks on tropical lep species are well received.

In the 2014 winter edition of the NABA magazine, *American Butterflies*, Glassberg took Satterfield and the *A. curassavica* naysayers to task.[8] He pointed out that nonmigrating monarch butterflies in Hawaii and Arizona have less-than-average levels of OE infection and that some monarchs overwinter and sustain themselves on evergreen milkweeds like fringed twinevine and pineneedle milkweed. He went on to suggest that high levels of OE infection associated with tropical milkweeds are caused by temperature effects or other factors that have nothing to do with the particular species of milkweed commercial plant growers have chosen to make widely available. Had commercial growers chosen swamp milkweed, *A. incarnata*, a relatively easy-to-grow pink-blooming milkweed and one that is native and has a broad range, would we be vilifying swamp milkweed? *A. incarnata* is not as resilient as *A. curassavica*, but with selective breeding it could surely come close to matching the consistent regeneration and ease of growth that made the latter a FleuroStar award winner.

Range and nativity are moving targets in today's context of global climate change. In 2012 and 2023, the USDA redefined hardiness zones in response to warmer weather patterns.[9] The pastel-shaded grow zone maps on the backs of seed packets that have guided gardeners for decades on when and where to plant certain species changed forever. San Antonio moved from Zone 8b, with annual lows of 15–20 degrees, to Zone 9a, with annual lows of 20–25 degrees—the same planting zone as coastal cities Houston and Corpus Christi. The new map reflects thirty years of temperature data, excluding the most recent years of global warming drama.

Timing is imperative, too. As grow zones change, the plants often can't keep up. Monarch butterflies head north to Texas in March. In a cool spring, native milkweed plants are not yet out of the ground as the furtive female butterflies fly north in search of a place to oviposit the first generation of eggs. Many San Antonio springtimes find native milkweeds still dormant while robust, reliable tropical milkweed from the previous year stands at the ready in San Antonio gardens, poised to receive the fertile bounty that would hatch the season's first generation of monarch butterflies. Plenty of fine, tender new leaves sprout from stems cut to the ground the previous fall, as per Satterfield and Monarch Joint Venture's recommended best practices. They await the hungry critters that hatch in the days after the migration's early arrival. Had tropical milkweed been unavailable in San Antonio gardens, the migrating female monarchs laden with eggs fertilized earlier in their travels would have to keep flying. Who knows where or if they would find a place to lay eggs before perishing? Chances of milkweed in early spring decrease the farther north one travels.

Glassberg summed up the quandary, professing NABA's overwhelming preference for native plants but stressing that tropical milkweed can serve as a "life buoy" for migrating monarch butterflies during a time in their life cycle when nothing else is available. Sometimes nonnative plants fill a butterfly need, Glassberg wrote, adding that "demonizing non-native plants is a misguided strategy that antagonizes many people who would be natural allies in our mission to conserve butterflies."[10]

Commercial butterfly breeders make up another constituency that considers the vilification of tropical milkweed overwrought. Commercial breeders rely heavily on tropical milkweed and, in recent years, the even more exotic *Calotropis gigantea*. The African native does not belong in the *Asclepias* genus, but it is a closely related cousin that belongs to another family of two hundred species. Butterfly breeders love the giant milkweed, as it's known, since the plant grows six feet tall and has enormous leaves that provide ample sustenance for hungry caterpillars—the perfect "cat food." *Calotropis gigantea* also exhibits high toxicity and flaunts dramatic cylindrical seed pods covered with light green bristles. The plant sometimes goes by the name balloon plant or hairy balls milkweed.

Tropical milkweed fans liken OE in monarch butterflies to the *Staphylococcus* bacteria in humans—present in a percentage of the population but causing debilitating symptoms only under stressed conditions. And, indeed, OE is present in many monarch butterflies. Some believe it is simply a part of the evolutionary cycle, killing off butterflies less fit than others. "If it were as deadly as many people imply, there wouldn't be an OE issue," suggested butterfly breeder Edith Smith of Shady Oak Butterfly Farm.[11] Smith has raised hundreds of thousands of OE-free monarchs for education, celebration, and research purposes. She and others suggest that OE is just one more Darwinian check that nature employs to keep the monarch population balanced.

As noted earlier, several studies have indicated that tropical milkweed, with its resilient habit, appealing chemical constitution, and presence in a warming climate, induces migrating monarchs to break their diapause and become reproductive, thus threatening the monarch butterfly migration.[12] The availability of nonnative milkweed species can also impact monarch wing development to the point that the creatures are less fit for migration.[13] And, of course, the continued presence of resilient tropical milkweed serves as a hotbed for diseases like OE.

Chip Taylor, who started an online milkweed market in 2010 to make native milkweeds more available to the public, expressed exasperation with the focus on tropical milkweed. "Just cut the dang stuff down," he famously said. Taylor sees urban development, climate change, and the overuse of genetically modified and herbicide-tolerant crops as threats worthy of focus. These realities have removed millions of acres of milkweed from our habitat each year. Taylor views the focus on tropical milkweed as a misplaced priority and "trivial issue."

Though native varieties are starting to become available for gardeners aiming to support the butterfly population, most have little choice when it comes to buying milkweed plants. On the rare occasions that native milkweed seedlings are available—at pop-up plant sales and specialty nurseries—transplanting often results in failure.

Gardeners and ecological restoration specialists bemoan native milkweeds persnickety ways. Many native milkweeds have a long taproot that makes potting and transplanting a challenge. Growing native milkweed from seed can also be complicated and often requires seed stratification, moist conditions for forty-five days, specific soil conditions, and alternate dry and wet periods. It can take two years to grow a healthy plant with enough "wow factor" to justify its cost. And since milkweeds are typically an ecosystem unto themselves, attracting aphids and milkweed beetles like a magnet, growers must jump through extra hoops to make sure the plants remain attractive enough to invite buyers. Systemic pesticides, which would knock out any pests, are verboten when producing caterpillar food.

Native milkweeds' fussy demands make for a tough case when commercial growers are scouting their next FleuroStar plant. In addition, commercial growers often must compete with "free" seeds and subsidized native plants. Milkweed and monarch evangelists are working hard to make the natives available by subsidizing their production through grants and subsidies.

Over at the Milkweed Patch in San Antonio, we've seen big changes in recent years. The homogenous butterfly haven of tropical milkweed that flourished in 2012 and attracted the attention of scientist Satterfield and monarch butterfly lovers no longer exists. Last time I visited, numerous nectar species—lantanas, purple mistflower, passionflower vine, and others—now create the plant palate. The City of San Antonio has not ceased installing milkweed patches along the San Antonio River Walk, however. There, the reliable perennial draws monarchs and other butterflies nine months of the year, keeping the tourists happy.

The gardening quandary calls to mind the locavore food movement: idealistic, admirable, but not always practical—or even possible. Imagine driving cross-country with your family and everyone is famished. Sure, you'd prefer to stop at a regional eatery where good food is whipped up from scratch from local organic ingredients, responsibly harvested, lovingly prepared, delicious, nutritious, and affordable. But that's not always possible. Sometimes you have to hit the drive-through of a fast-food joint because that's all there is. And that gets you to the next place.

▼ ▼ ▼

CHAPTER 13

THANKS, MONSANTO!

My Monsanto salad arrived in a brown paper bag. Its dried red cranberries, blue cheese crumbles, and caramelized pecans were welcome additions to the predictable romaine mix. Before my visit to the international agriculture company's headquarters, Pamela Bachman, a science fellow who works closely with the monarch butterfly community, had thoughtfully asked me my preferences. On this sunny spring day in 2017 just months before the announcement of Bayer Corporation of Germany's acquisition of the company, a select group of Monsanto staff joined us for preordered sack lunches in a conference room at their 220-acre research facility in Chesterfield, Missouri.

Is it true that Monsanto bans genetically modified foods in its cafeteria because the company "believes in choice" for its employees, I asked the group assembled for my benefit after a tour of the agrichemical giant. The notion was first floated in the late 1990s when a British company that supplied catering services to a Monsanto pharmaceutical factory in Buckinghamshire, England, posted a notice in the company cafeteria declaring its rejection of genetically modified food,[1] and it has reverberated ever since.

No, Bachman said cheerfully. Just check "The Top 12 Myths about Monsanto" handout shared earlier in the day.[2] Later, I reviewed the one-sheeter. The document tackled the usual claims: Foods produced from Monsanto's genetically modified crops are unsafe. Monsanto products kill bees and other pollinators. Monsanto's Roundup causes cancer. Monsanto's industrial agriculture model is unsustainable. And number one on the list: Monsanto employees don't eat GM food (they only eat organic foods) in our cafeteria.

Each "myth" was countered by a curated set of facts, articulated in deft language, presented as truth. The "truth," according to the document, is that all types of foods can be found in Monsanto cafeterias. And, by the way, Monsanto holds a special event each summer at which employees can purchase genetically modified sweet corn from local farmers. "It sells out quickly," said Bachman.

Bachman and the professional communicators at Monsanto are well schooled in debunking these myths and addressing the apparent hypocrisies inherent in their jobs. And they seemed genuinely sincere when describing their personal and corporate commitments to monarch butterfly and pollinator conservation.

A self-described tree-hugger, Bachman earned her PhD in ecotoxicology at the University of Florida and worked for years in environmental research and academia before joining Monsanto in 2008. Today she wore longish, blonde-brown hair and an easy smile and shared that she'd been working to reduce lawns in her neighborhood, planting native and pollinator plants, and encouraging neighbors to do the same. Monsanto employees have a company bee club, she said with pride, and the group stages honey tastings and pollinator "happy days" for staff. Another team member chimed in that up in Williamsburg, Iowa, Monsanto employees set up a monarch butterfly hatching tent the previous fall. They raised, tagged, and released monarchs during the migration. "It makes you feel good," said Bachman.

Earlier that morning, we gathered in a different conference room in which Bachman had laid out Monsanto's monarch butterfly conservation and communications strategy in a PowerPoint presentation. Its goals: align interests, lead by example, be a convener, and create awareness and action. This strategy, in place since 2015, put Monsanto in the middle of the growing monarch butterfly conservation movement—as both a benefactor and a villain.

As a benefactor, Monsanto allocated millions of dollars to monarch butterfly conservation early on. From 2015 to 2017, in the wake of President Obama's National Pollinator Strategy, the company awarded at least $4.4 million in cash to organizations conducting monarch butterfly conservation. The bulk of that financial commitment, $1.2 million annually for three years, passed through the National Fish and Wildlife Federation (NFWF), a 501C-3 created by Congress in 1984 that has grown to become one of the nation's largest conservation grant-makers.[3]

By the end of 2017, $3.6 million dollars from Monsanto had been deposited in the NFWF's newly created Monarch Butterfly and Pollinators Conservation Fund (MBPCF). The separate money bucket was set up in early 2015, just months after the petition to list the monarch as "threatened" under the Endangered Species Act was presented to the Department of the Interior. Monsanto said it also dedicated another $800,000 to select monarch butterfly conservation groups over the three-year period. Monarch Watch, Pheasants Forever, the University of Guelph in Canada, Missourians for Monarchs Collaborative, the Iowa Monarch Conservation Consortium, Sand County Foundation, and the Keystone Policy Center have all received Monsanto money.

At the time, the Monsanto NFWF investment constituted almost a third of the total $11.1 million that made up the MBPCF. Those funds had been matched and leveraged by other contributions. Partnerships with government agencies—U.S. Fish and Wildlife Service, USDA, U.S. Forest Service, U.S. Geological Survey, and others—also multiplied the gift.

The Monsanto grants applied to the MBCPF directly or indirectly fueled every major pollinator conservation effort in the country. Monarch Joint Venture, Pollinator Partnership, Texan by Nature, U.S, Fish and Wildlife Service, North American Butterfly Association, Texas Parks and Wildlife Federation, and others have all received grants from the fund. Even the Xerces Society—the Portland, Oregon, invertebrate advocacy association that launched the petition to list monarchs and is known for its purist bent—received Monsanto-supplemented NFWF funds. And Mexico City–based Milenio, which has covered the monarch butterfly story extensively, cites Monsanto as a sponsor of its eleven-part series and subsequent documentary, *The Monarch: Spirit of the Forest*.[4]

In its MBPCF program report issued in early 2018, the NFWF cited seventy projects funded, 730 workshops or meetings hosted, 13,200 pounds of native milkweed and seeds collected, 127,000 acres restored or enhanced, and 730,000 native milkweed and other forb species propagated. Grantees matched this investment with in-kind, staff, or cash resources of $18.1 million, for an impact of more than $29.2 million, according to NFWF.

By early 2023, according to the MBPCF website, the fund had awarded $19.7 million to 123 projects since 2015. "Grantees have matched this investment with an additional $32.3 million for a total on-the-ground impact of $52 million." The website goes on to cite the following accomplishments: restoration or enhancement of 352,800 acres, propagation of 1.1 million native milkweed seedlings, collection of 3,000 pounds of native milkweed seeds, and coordination of 1,600 workshops and meetings.

We can thank Monsanto/Bayer for the dozens of useful webinars staged by Monarch Joint Venture and made available for free to monarch conservationists in recent years. And those hard-to-find, local ecotype milkweeds, free to schools, community groups, and large restoration efforts grown by private nurseries and distributed through Monarch Watch's Bring Back the Monarchs campaign—Monsanto funding made the program possible. Monsanto/Bayer dollars also helped bankroll the myriad milkweed studies and censuses conducted along the Interstate Highway 35 corridor that continue to shape monarch and pollinator conservation policy.

Funneling the money through the NFWF fund leverages Monsanto's resources and, to some extent, launders the money for the outside world, making it appear less like a deal with the devil. But perhaps even more significant are the untraceable, indirect Monsanto resources and affiliations that influence the course of pollinator conservation.

During our conversation at Monsanto headquarters, Bachman shared that she sat on the granting committee of the NFWF monarch butterfly and conservation fund. She helped determine which grants are approved and which are rejected. The U.S. Fish and Wildlife service was "in step" with Monsanto, said Bachman during our visit. She and comrade Aimee Hood, another scientist communicator, were "deeply involved" in state monarch conservation initiatives. Monarch Joint Venture staff "unofficially" vetted Monsanto outreach materials. Graphic designers commissioned by Monsanto created web pages, logos, and collateral materials that were printed and paid for by Monsanto.

And then there's the swag. Monsanto staff set up booths at the large agricultural trade shows, working closely with farmer-oriented trade groups like the National Corn Growers Association and the American Soybean Association, spreading the gospel of pollinator advocacy through attractive Monsanto-branded, monarch-themed giveaways. How about a "Farmers for Monarchs" luggage tag? Or a pollinator habitat restoration kit, packaged in a giftlike cardboard box? It includes a pair of high-quality gardening gloves, milkweed seeds, a monarch fun facts info card and bookmark, as well as a fancy Monsanto branded rain gauge. "Because who doesn't love a rain gauge?" asked Bachman. (Disclosure: I put one in my pollinator garden.) Bachman openly conceded that at these trade shows people don't want to talk to Monsanto. "So we make them talk to us so they can get the swag."

Though it's impossible to put a definitive price tag on Monsanto's total investment in monarch butterfly and pollinator conservation, by all appearances and measures the company has committed more financial, in-kind, and human resources to the cause than any other single corporate entity.

One of the communications team's most apparent points of pride is Monsanto's Farmers for Monarchs initiative. "It's just a great story to tell," said Bachman. She, Hood, and head of corporate strategy Tracey Reynolds hatched a plan to engage farmers to utilize less fertile cropland on behalf of biodiversity. They did not act in isolation. "You have Monsanto, you have Bayer, BASF . . . all doing separate things," said Bachman. The agrochemical colleagues were all telling the story of monarch butterfly conservation in a different way when the Monsanto team realized that the community would be confused, seeing different messaging coming from different sources. "If you want to get a movement and you want to get things done, you have to have consistent messaging across the board," she said.

That consistent message, "stems in the ground"—milkweed stems, to be exact—has driven communications by Monsanto, nonprofit organizations, government agencies, and now the farmers and trade associations, since 2015. It speaks to Monsanto's ongoing strategic communications goal of creating "aligned interest."

On the villain front, Monsanto holds the distinction, more than any other agribusiness, of playing a leading role in the decline of the monarch butterfly migration. The company's introduction and effective proselytizing of Roundup Ready crops have decimated the monarchs' host plant in the insects' crucial midwestern summer breeding grounds. Throughout most of the twentieth century, common milkweed, *Asclepias syriaca*, coexisted with the agriculture fields of the pastoral American farm belt. The robust, rhizome-spreading milkweed with stiff stems, large pink umbels, and big oblong seed pods could frequently be spotted on the edges and between the cornrows of agricultural land. Common milkweed developed a reputation as hearty and sometimes overly gregarious, but seldom did it disrupt crop production, as did other weedy perennials and grasses. It thrived on disturbance. Thirty years ago, reproductive monarch butterflies traversing the midwestern heartland during late spring and early summer found plenty of available host plant on which to lay their eggs.

But then in 1994, Monsanto gained USDA approval for its glyphosate-tolerant soybeans. The soybean seeds had been genetically modified to tolerate glyphosate, Monsanto's popular broad-spectrum herbicide marketed as Roundup. Roundup is nonselective; it indiscriminately kills most plants. It prevents the uptake of certain proteins that allow the plant to grow. Monsanto scientists identified genes in common soil bacteria found in a Louisiana production plant's sludge pond that were resistant to glyphosate. They spliced this gene, known as CP4, into select crop plants. The result: the genetically modified seeds produced crops that can be sprayed repeatedly with glyphosate with no harm done to the plant while killing all other weeds and foliage with which it makes contact. This "system" of marketing genetically engineered seed with its nonselective herbicide soul mate became a signature moneymaker for Monsanto.

Within two years, genetically modified corn had entered the landscape. By the year 2000, genetically modified cotton and canola had joined the Monsanto system lineup—and roughly half of the company's $5.5 billion in sales came from glyphosate. Monsanto told shareholders that glyphosate product sales jumped 18 percent in the previous twelve months.[5] By 2011 an estimated 94 percent of the soybeans, 72 percent of corn, and 96 percent of cotton in the Midwest was planted in herbicide-tolerant varieties.[6] The amount of glyphosate used annually on the landscape vaulted from 10 million pounds in 1995, prior to the introduction of glyphosate-tolerant soybeans, to 205 million pounds in 2013, a dramatic twentyfold increase in just eighteen years.[7] Most disturbing, scientists and monarch conservationists agreed, is that more Roundup was being used on more land, more frequently and later in the season, because the weeds were evolving to resist glyphosate. The phenomenon is known as the herbicide or pesticide treadmill, the need to ratchet up the use of herbicides or pesticides continually as weeds and insects develop resistance.

This has been terrible news for common milkweed, wildflowers, and other volunteer forbs that once populated the agricultural fields of the Midwest. Because the Roundup Ready system dominated land management of the monarchs' primary summer breeding and feeding grounds, the space between the crop rows became a sterile wasteland. Only corn, soybean, wheat, or other genetically modified crops that could take the serial blasts of the most common herbicide in the world remained. Common milkweed became much less common.

A plant survey conducted in 1999 found that low densities of common milkweed were present in approximately 50 percent of Iowa corn and soybean fields. Ten years later, common milkweed was documented in only 8 percent of surveyed fields. And the cultivated fields in which common milkweed grew incidentally were reduced by approximately 90 percent in 2009 compared with that witnessed in 1999.[8]

It's no coincidence that the folks assigned to serve on Monsanto's monarch and pollinator conservation outreach team worked closely with the company's regulatory personnel—the same troupe of Monsanto employees who monitor regulatory issues like the potential listing of monarchs as threatened under the ESA. Nor was the timing of Monsanto's woke monarch conservation ethos a fluke.

The "stems in the ground" mantra began in 2015, shortly after the ESA petition was filed. Should the monarch butterfly become listed as a threatened species, the bottom lines of Monsanto and other chemical giants would take a serious hit. Such a turn would cast milkweed as "critical habitat" as defined by the ESA. Destroying milkweed would become a crime punishable by fines or mitigation. Civil penalties of $25,000 and criminal fines up to $100,000, as well as prison sentences, could be assigned for the "taking" of milkweed, the ESA verbiage for messing with or destroying a threatened species' habitat.

As we started on our lunches, talk turned to the esoteric and often contentious issues that divide the monarch butterfly community. How did Monsanto feel about tropical milkweed? They adopted the Monarch Joint Venture company line. Okay to plant it, just cut it back in the winter in warmer climates.

What about commercially reared butterfly releases? The group was unfamiliar with the topic. I explained the concerns and ongoing debate and mentioned that, while definitive proof has not emerged that releasing commercially raised butterflies into the population harms the wild butterflies or the ecosystem, some scientists consider adding monarchs bred in captive, for-profit circumstances to the wild a terrible idea.

Monsanto scientist and communications lead Aimee Hood, who worked closely with farmers and the bee community, was curious. She turned to her peers. "What does our science think?" she asked.

"*Our* science?"

Conflicting science is nothing new. Those who study *Danaus plexippus* often disagree about priorities and approaches to monarch conservation and frequently cite opposing studies, the absence of data, and flawed scientific models that conflict with their point of view. "The science is ongoing" is an oft-cited refrain. But Monsanto's penchant for preferred facts is well documented and powerfully effective in driving the company's agenda.

William Reeves, glyphosate expert, PhD, and regulatory policy and scientific affairs manager, joined us for part of my Monsanto tour. He professed zero concerns about the safety of glyphosate: "I use it all over the place. I have a little wand sprayer and I go dancing around the milkweed."

The discussion turned to author and activist Carey Gillam's 2017 book, *Whitewash: The Story of a Weed Killer, Cancer and the Corruption of Science.* Gillam cites dozens of examples of Monsanto-funded science in conflict with independent research. She devotes an entire chapter to "Spinning the Science." In it, she details the "hidden corporate hand" that funds presumably independent research through quiet money, front organizations, and unusual sponsorship arrangements.

When the European Parliament was debating the safety of glyphosate in 2016, the environmentally friendly Green Party suggested that all 751 Parliament members submit their urine for testing in an attempt to understand the extent of glyphosate in the human body. Of the forty-eight members of Parliament from thirteen countries who agreed to what became known as "the pee test," all forty-eight tested positive for glyphosate. Their urine samples reflected an average of 1.7 micrograms of glyphosate per liter, roughly seventeen times the level permitted in European drinking water.[9]

"It's urine," said Reeves, dismissing the issue. Urine is more concentrated than water, he added. Earlier in our conversation, he shared stories of gardening with his daughter. Did he have concerns about glyphosate in his daughter's blood or body? "I don't," he said emphatically. Then he retrieved what sounded like an old standby example from his pocket of relatable analogies: "Let's look at another registered pesticide: vitamin D." It's a rodenticide. Apparently vitamin D tablets, crushed into a powder, can be used to take care of any mouse problem that presents itself.

"You take the pill because your doctor says you have a vitamin D deficiency. You go outside, and it's uncontrollably being created in your skin—but it's also a rodenticide at a high enough dose. And that's really what it comes down to, is the dose you're getting. . . . As a toxicologist, I look at that and think, yeah, everything can do that sort of stuff."

As I drove through the Monsanto security gate for my tour that morning, flashbacks from my days as an account manager at Business Wire visiting publicly traded companies crowded my thoughts. The routine was familiar. Pass the security checkpoint, sign in at the front desk, shake hands and make cheery small talk in the lobby. Wait for your contact to retrieve you and shepherd you to their office. Back then, I was selling something. Now, I was being sold something.

The tour of Monsanto was impressive. A quick trip to a machine-filled workroom brought us to the "corn chipper," a contraption that decapitates the heads of corn kernels, guillotine style. No photos allowed in this proprietary enclave, for this was the apparatus Monsanto developed to inspect and eventually insert genetic material into seeds. A hike across the courtyard and up the stairs took us to a floor of monitored grow houses where staff in white suits wore latex gloves and checked the status of soybean and corn experiments. Then we visited Climate Corp., which Monsanto acquired in 2013.

The technology subsidiary allows farmers to plot their land and assess temperature variables and geographic traits on an app that not only sells them crop insurance but also gathers data in the interest of "precision agriculture." The technology-infused farming uses control systems, sensors, robotics, GPS-based soil sampling, automated hardware, and software in a quest to gain more efficient agricultural outcomes. Precision ag can be useful in planning all agricultural endeavors—from determining which field is better suited to soybeans to which is more appropriate for pollinator habitats. "And the farmer owns the data," the tour guide pointed out.

As a longtime media and marketing pro, I was impressed with the comprehensiveness of Monsanto's reach and the zealous convictions of its staff, which rivaled those of comments made by members of the Monarchy on social media. In just three years, and for only a couple million dollars, they had assumed control of the conversation with their "stems in the ground" message, infiltrating every major constituency in the monarch world. Academia, nonprofits, the farmers, the trade associations, politicians, the EPA and USDA—all touched by their influence and within their realm or grasp.

"We're not always on the same side, but for this we do have aligned interest," said Bachman, alluding to her colleague agrochemical companies, Dow Chemical, Syngenta, BASF, and their future owner, Bayer, all partners in this influence quest. "You get that critical mass of groups together, and you have the power to make change. Somebody had to be the voice. . . . We kind of had the opportunity to do that. Somebody had to do that and I think it's really a special thing."

In June 2018, Bayer AG, the German drug and chemical company, acquired Monsanto in a $66 billion merger. One of Bayer's first moves was to drop the Monsanto name, converting the St. Louis headquarters to the North American Crop Science division of Bayer. In addition to inventing both aspirin and heroin, Bayer has a history in Europe of bee conservation, Monsanto representatives said at our lunch. They suggested that in the future the U.S. conservation focus on monarch butterflies would continue through "synergistic efforts."

Not long after Bayer's acquisition of Monsanto, the first Roundup lawsuit was decided. A California jury ruled unanimously that the company pay groundskeeper Dewayne "Lee" Johnson $289 million in damages for the role its herbicides, Roundup and Ranger Pro, played in causing his cancer, a non-Hodgkin's lymphoma.[10] Johnson worked for years as a groundskeeper at a California school. Bayer appealed the verdict, and a judge reduced the award later that year. Finally, in 2021, Johnson received a $20.5 million payout, a fraction of the initial amount the jury awarded him in 2018.

In the years since our initial meeting, Bayer has endured numerous lawsuits accusing Monsanto—now recast as Bayer Crop Science—of knowingly selling Roundup and other glyphosate products that they were aware caused cancers. According to Forbes, as of May 2022, Bayer/Monsanto had settled more than 100,000 Roundup lawsuits, paying out about $11 billion. In 2023, a Missouri jury ordered the company to pay more than $1.5 billion to three former Roundup users, all of whom developed cancers, in what was labeled one of the largest trial losses in the five-year litigation over the herbicide.[11] The herbicide has been banned in at least ten countries (including Germany, where its owner is based), and its use is restricted in select cities and states. In New York City, for example, a group of kindergarteners lobbied their elected officials for years to ban toxic chemicals like Roundup in parks, playgrounds, and other public spaces. The pesticide ban was approved in April 2021.[12]

Monsanto's new owners may be experiencing deja vu as lawsuits and product bans begin to pile up on another class of chemicals decimating insect populations—neonicotinoids. Bayer was the largest producer of neonicotinoids on the planet when it was introduced in the 1990s but now has about 20 percent of the market.[13] Neonics have been banned in Europe since 2018.

▼

By 2023, Bachman had become an associate science fellow at Bayer's Climate Corporation. At the separately branded company under the Bayer umbrella, she manages a team of six in support of FieldView, a digital farming platform that works to "get farmers to think more holistically about the resiliency of their farms."[14]

Tim Fredricks, environmental engagement manager, has assumed Bachman's role as outreach specialist. Like his predecessor, he holds a PhD in environmental toxicology and now sits on the granting committee of the MBPCF. According to Fredricks, Bayer is taking "a more holistic approach" to conservation, and he points to the company's investment in HabiTally, a free app developed by Bayer and its wholly owned subsidiary Climate Corp and the University of Iowa as an example. The application allows landowners and private citizens to support monarch butterfly recovery by registering their habitats and entering data about the types of plants installed as well as sightings of monarch butterfly activity in all its stages.

According to Fredricks, HabiTally has 1,800 habitats registered and even enables milkweed stem counts through drone imagery. "It allows us to count habitat that isn't part of a government program that won't get counted in any other way," said Fredricks. The company's new conservation tagline, "Bayer for Biodiversity," says it best, he asserts. And as a beekeeper, Fredricks believes that "without habitat, no matter what else is happening won't matter."

Fredricks said Bayer is funding conservation at or above previous levels. In their Monarch Flyer, posted online, the company touted more than $21.4 million invested in nonprofit monarch butterfly and pollinator conservation efforts. Fredricks said the company has contributed another $5.1 million to the MBCPF during that time, bringing the total since 2015 to about $26.5 million.

Depending on where you sit, Monsanto/Bayer's role in monarch and pollinator conservation can be viewed as greenwashing, conservation mitigation, or Faustian bargain. Many would argue that Monsanto's early millions in monarch conservation seed money and continued track record of several million dollars per year constitute paltry recompense for a company that has played such a significant role in monarch butterfly and pollinator habitat decline. In the first three years that Monsanto committed more than $4 million to monarchs, the company enjoyed a $23.1 billion net profit.[15] In 2021, Bayer, its new parent, saddled with lawsuit settlements but persistent growth, realized profits of more than $7.2 billion.[16] Financials for 2022 showed the company with €2.3 billion in profits–about $2.5 billion.

CHAPTER 14

▼ ▼ ▼

AVOCADO WARS

Attendees described a meeting at the El Rosario butterfly sanctuary on January 13, 2020, as unremarkable except for one thing: Homero Gómez González's phone was blowing up. Bystanders couldn't help but notice he was getting *lots* of phone calls.

According to news reports, Gómez, an international butterfly hero and the manager of El Rosario, finally picked up the phone.[1] Someone was lobbying him to come to El Soldado de Ocampo, where the final day of a public fair was unfolding. "Yes, yes, of course I'm going," he was overheard saying to the caller. Soon, the stocky, mustached fifty-year-old set out for the fair, about a half-hour drive away.

Gómez arrived at the fair in the late afternoon. He visited with locals and elected officials, including local politicians who said he was drinking, dancing, laughing, and schmoozing until he left at about 8 P.M. According to his family, Gómez was spotted at the fair until about 9:30. It was the last time he was seen alive.[2]

When he didn't return home that night, his wife Rebeca was worried. The next day she reported him missing to the police, and a search party formed. Hundreds of residents joined local and state authorities looking for "Señor Homero," as they called the butterfly activist. Gómez managed Mexico's most visited butterfly sanctuary and invited people to visit by posting enchanting photos and videos of butterfly masses at El Rosario on Facebook and Twitter. In what appears to be one of his final Facebook posts, the affable activist stands before a cloud of flitting butterflies, his arms spread wide in a welcoming gesture. "Come and see this marvel of nature," he says, talking directly to the camera. "They're the souls of the dead, lovers of the sun, simply our world heritage. We await you."

For weeks the search parties found nothing. Then, on January 29, Gómez's body was found in a well about 150 meters from where the fair took place—a location volunteers and others had previously searched but found nothing. He was clothed in the white guayabera and gray suit pants he was seen wearing at the fair. Authorities said he had likely been dead for two weeks, the same amount of time he'd been missing. The cause of death was determined to be asphyxiation and head trauma.

A few days later, part-time sanctuary guide Raul Hernández, forty-four years old, also turned up dead. His death was ruled a likely homicide from a deep blow to the head from a sharp object. Officials have been unable to establish a connection between either the two deaths or the two victims' work in monarch butterfly and forest conservation. But others attribute the incidents, especially the death of Gómez, to his outspoken and proactive stance in protecting the butterflies and the forest. Like many of his fellow *ejidatarios*—the community members who share responsibility and ownership of communal lands—Gómez grew up as a participant in the logging industry. As the eldest of nine children in a logging family, he started out harvesting the forest. And when the government first announced the formation of the Monarch Butterfly Biosphere Reserve in his corner of Michoacán, he opposed it.

But after the Mexican government set aside 62 square miles for monarch sanctuaries under the banner of the Monarch Butterfly Biosphere Reserve (Mariposa Monarca Biosphere Reserve) in 1986, followed by an expansion to 217 square miles in 2000, Gómez came around. He understood that tourists coming to see the butterflies could generate jobs and revenues for the community. His career at El Rosario had him evangelizing for butterflies and the forest that hosted them for years.

So what happened to Gómez and Hernández?

"Follow the avocados," David Bohlken, IBBA president wrote me in an email in March 2019. "It's all about the avocados." Bohlken's words resonated ten months after reading them when I first learned of Gómez's death via the news coverage that followed. The Oklahoma butterfly breeder and partner in the Native American Euchee Farm had shared a wild story with me in conversations we'd been having about the commercial breeding business during the research of this book. In a series of emails, I asked Bohlken if he saw commercial breeding playing a role in monarch butterfly and pollinator conservation beyond producing livestock for weddings, funerals, other events, and exhibits.

He responded that, yes, in fact the IBBA had been working closely with the Mexican government to establish an "emergency breeding program" in the event of a catastrophic ice storm or other extreme weather event hitting the sanctuaries someday. "Five million butterflies would be the goal," he said.

The idea was a shocking notion given scientists' and conservationists' general resistance to commercial butterfly breeding and releases, not to mention geographic appropriateness, respect for native species and ecosystems, and human meddling in a natural wonder designated a World Heritage site.

Bohlken shared that he'd been approached by Mexican government officials at the twenty-fifth annual meeting of the Commission on Environmental Cooperation (CEC), a trinational, intergovernmental conservation organization representing Canada, the United States, and Mexico that was born out of NAFTA.[3] The meeting took place in June 2018 in Oklahoma City not far from the butterfly farm Bohlken operated with his wife Jane Brackenridge in Bixby.

Preliminary plans for releasing commercially reared monarch butterflies at the roosting sites in the event of a catastrophic storm were in motion when Andrés Manuel López Obrador was elected president of Mexico a month after the CEC meeting. According to Bohlken, once AMLO, as he's known, was elected, the plan was scrapped. "Everyone I worked with was replaced with new people from the new administration—and they are all pro-avocado people. And that is what's going to destroy the monarch butterfly," wrote Bohlken.

▼

A pro-avocado stance has existed for more than 1,500 years in Mexico, where the delicious, healthy fruit has been cultivated since 500 A.D. The Aztecs believed avocados resembled testicles and assumed that consuming them increased one's strength and served as an aphrodisiac. When Spanish conquistadors first encountered the large, oblong berry in the early sixteenth century, they called it *aguacate*, after *ahuacatl*, the Aztec word for testicle.[4]

Cultivated for millennia in Central America, avocados entered the U.S. market only in the late nineteenth century when Judge R. B. Ord of Santa Barbara, California, brought in a tree from Mexico. Growers experimented with different varieties for decades, but few had commercial success until Rudolph Hass patented the Hass avocado in 1935. The richer taste, longer shelf life, and higher yields made the Hass variety a hit. For years, California supplied most of the avocados in the states, and the seasonal offering typically sold out and cost more than most fruits and berries in the produce aisle.

All that changed with the signing of NAFTA (updated as the USMCA in 2020) in 1994. The treaty paved the road for avocados to be imported from Mexico for the first time since 1913, when Mexican avocados were banned for U.S. import because of phytosanitary concerns such as the possibility of pests and diseases that might infiltrate local avocado markets in California and Florida.

The concerns about plant health led to the establishment of strict import guidelines, and finally in 1997 Michoacán was the only Mexican state authorized to export avocados to the United States. In the summer of 2022, the state of Jalisco was also authorized to import avocados. When NAFTA was signed in 1994, Americans consumed about one pound of avocados per person per year. In 2021 they consumed 8.43 pounds per capita, making avocados a $3 billion per year industry, 80 percent of which was supplied by Mexico.[5]

That massive profit potential has earned avocados the moniker in Mexico *oro verde*, or "green gold." A less flattering label: "blood fruit." Because of its lucrative nature, avocados have attracted unsolicited interest from Mexican crime cartels that view avocado farms as their next-best revenue stream. And they're not shy about using intimidation, extortion, tariffs, farm seizures, "disappearances," and murders of growers to claim it.

News accounts have described continuing encroachment of cartels on Mexico's more than 34,000 avocado growers and forty-four packers–the vast majority of whom run small family farms of less than five acres.[6] A disturbing forty-three-minute video documentary by the Canadian news magazine W5, "Narco Avocados: Violent Drug Cartels Are Taking Over Mexico's Avocado Industry," demonstrated the terror witnessed in what has become known as Mexico's "Avocado Strip."[7] The Trans-Mexican Volcanic Belt boasts not only the ideal climate and soil conditions for overwintering monarch butterflies but also the perfect setting for high-yield avocado production year-round, with the help of supplemental water and regular doses of pesticides and herbicides.

W5 correspondent Avery Haines dug into the details, interviewing avocado farming vigilantes and cartel bosses alike. The report in early 2023 starts out on an avocado farm with "Gabriel," whose real name and identity are omitted for his own safety. Haines describes him as "armed to the teeth" and points out that he is forced to pay one cartel for protection from another while also defending his family and farm with his personal AR-15 assault weapon.

"If you're doing well, they'll kill you or disappear you. . . . they'll pull you and your family out of your house and take away your orchard . . . and take all the money," Gabriel says through a translator. He describes how the cartels seize land: "They just say, 'I like your land. I like your house. You have to vacate it, you have 24 hours to leave' and if you don't, they come and kill you."

The report recalls earlier accounts of brutal cartel intimidation, such as occurred in 2006 when gunmen associated with the La Michoacana cartel tossed five human heads on the dance floor of a bar in Uruapan. Or in 2019 when the Jalisco New Generation cartel mutilated nineteen people—and hung nine of the seminaked corpses from a bridge in a city west of the capital. At least ten other dismembered and bullet-riddled bodies were reportedly found dumped in two nearby locations.

In 2021, "Martín," one of many avocado farmers subjected to extortion, told a reporter he was forced to pay las malosos—Spanish for "the bad guys"—a fee of 12,000 pesos (about $600 at the time) per hectare of avocado farm. "In exchange for their 'protection' they wanted to sell crystal [methamphetamine] in the communities, or they want to initiate avocado plantations in the upper hills where monarchs and many other species live," he said.[8]

In early 2023, an anonymous citizen submitted a petition to the CEC requesting that the organization evaluate a complaint. "Mexico is failing to effectively enforce its environmental laws to protect forest ecosystems and water quality from the adverse environmental impacts of avocado production in Michoacán, Mexico," it asserted. A July 2023 update on the CEC website, referring to the National Water Law and the General Law of Ecological Balance and Environmental Protection, indicated that "provisions characterized as environmental law are not germane to the assertions raised in the submission." This response went on to state that waters used for irrigation of avocados were not sourced from national waters or were tapped from "freely available groundwater. . . . In accordance with the provisions of Article 24.27(4)(a), Mexico requests that the CEC Secretariat proceed no further with submission SEM-23–002 (Avocado Production in Michoacán)."[9] In short: case dismissed.

So, should butterfly advocates and guacamole-loving Americans stop eating avocados? Forest ecologist Cuauhtémoc Sáenz-Romero, a man well acquainted with the forests and the butterflies that occupy them, says no, people should keep consuming them. Sáenz-Romero, who works at the Institute for Research on Natural Resources at the Michoacán University of San Nicolás de Hidalgo in Morelia, has studied the high-elevation oyamel fir forests where the monarch butterflies roost extensively. He acknowledges that avocado production taxes the ecosystem with its demands for scarce water and pesticide use. But, he points out, the industry serves as an important source of jobs and regional income—not only in Mexico but in the United States.

What could rural Mexicans do in the event of an effective boycott? he asked. "They might try to migrate to the U.S. and increase the current humanitarian tragedy that's going on." Sáenz-Romero suggests, instead, a system of independent and ecologically sustainable international certifications of "green" or "sustainable" avocado production, as happens today with timber and fair-trade coffee. He points to "green wood" programs established by the International Forest Stewardship Council already in place in Mexico's indigenous forest communities in Michoacán, Oaxaca, and elsewhere. The timber is certified only after the logging industry complies with a list of sustainability requirements.[10]

Like the general public who loves avocados and butterflies, monarch butterfly conservationists have mixed feelings on the topic. David James, an entomologist and monarch researcher at Washington State University who is focused on the California population, said that only "a tiny proportion" of Mexican avocados are grown near the monarch overwintering area. "Any avoidance of Mexican avocados would likely have no impact, unless you know exactly where they originated."

Migration studies expert Andy Davis stated up front that he loves avocados. He labeled it "unfortunate" that one of the best places in the world

to grow avocados also happens to be the same location of the monarch butterflies' overwintering sites. "There are poor farmers just trying to make a
living, then there are cartels encroaching into the business, and there are
guns and violence too. All of this is superimposed on the needs of the monarchs, which need these same high-elevation landscapes for three months of
the year," said Davis, who believes the North American monarch population is faring quite well. He doesn't see the harm in eating avocados despite
pressures on their winter quarters and the "doom and gloom often reported
in the media."

Longtime monarch researcher and conservationist Karen Oberhauser
is pretty passionate about the avocado issue, having watched the spread
of farms near the monarch overwintering sites for two and a half decades:
"You can't travel in the area without noticing avocado plantations creeping
higher and higher up the mountains." She added that, since the NAFTA
agreement, Mexican avocados have become much more affordable than those
produced in the states, a trend that is likely to continue now that Jalisco is
also exporting the popular fruit. "I remember when avocados were pretty
expensive, and you'd maybe buy one for a special occasion. But now they're
much cheaper."

Oberhauser said she no longer purchases avocados in the states, even in
restaurants, even though she loves their taste and consumes them in Mexico.
"It feels complicated to support this out-of-control growth in production that
has resulted in so much forest destruction. I realize I'm not going to solve
this problem through my personal actions, but it makes me feel like I'm at
least not contributing to the problem."

Columba González-Duarte, a sociocultural anthropologist and assistant
professor at Mount St. Vincent University in Halifax, Canada, argued that
the conservation approach of the Monarch Butterfly Biosphere Reserve and
adopted by the international community has encouraged organized crime
violence by removing the community from the forest. A self-described radical
who grew up in Mexico City, González casts the Reserve as "a frontier region of
unregulated economic activity in which violence redefines human relations."
Her research makes the case that the reconfigured authority and oversight
deepen the control of the cartels, facilitate the expansion of legitimate and
illegitimate economies, while undermining previous methods of sustainable

community forest management typified by a more "reciprocal" relationship with the forest. She adamantly considers the Reserve a failure. "The failure of this Reserve is also a failure of this model," she said in an interview.[11]

The issue is neither avocados nor monarchs but the disparity between two countries, says González, one that consumes carelessly and another that can't meet market demand ethically and sustainably. "I do not eat avocados when I am in Canada. Still, it's a cultural food for me, and if someone has the 'right' to eat avocados, it shall be Mexicans in their own country." She believes that "commodifying land" by paying community members to not log the forest turns what were once communally managed forests into a mostly human-free frontier resembling the wild west while facilitating organized crime groups and assigning responsibility for forest security to local residents. "Restrictions that have forced communities to abandon the core land's traditional uses have nurtured a cycle of declining community presence, granting space to organized crime in the reserve's core area."

González posits that the death of Homero González serves as a paradigm for the flawed approach of the Monarch Butterfly Biosphere Reserve. "As an old-style *ejido* leader, Homero first battled the reserve's creation because it dispossessed *ejidatarios*. Through the years, however, he changed views and found a place to keep leading the agrarian community by protecting monarchs and opposing illegal deforestation. . . . His disappearance illustrates how an *ejido* leader achieved adaptation to the neoliberal management of nature but was ultimately unable to thwart the expansion of the drug economy and its violence in the conservation area."

The friends and family of Homero Gómez González would likely agree. As of this writing, the case of his disappearance and death remains unsolved.

CHAPTER 15

MOVING THE FOREST

Cuauhtémoc Sáenz Romero wants to move the forest—the forest where the monarch butterflies roost. The bespectacled forest geneticist proposes relocating the forest of oyamel firs where millions of monarch butterflies roost each fall to a higher elevation to save it from climate change. By doing so, the monarchs will have a long-term winter home, the theory goes—if they find it, adopt it, and return to it. "We don't know if it will happen," the soft-spoken Sáenz Romero told a small group of visitors touring his experimental plot in 2017. He reminds us that those butterflies that do find their way here have never been to Mexico.[1]

Sáenz Romero's seemingly insane goal has an official name: it's called "assisted migration." The controversial approach to saving at-risk species involves the physical transfer of plants or animals to climates and environments more supportive of their growth than those in which they traditionally evolved—in short, a human intervention.

Sáenz Romero, his colleague environmental sciences PhD Arnulfo Blanco Garcia, and ecology students from the University of Michoacán Hidalgo in Morelia launched an assisted migration test site of the oyamel, *Abies religiosa*, commonly known as the sacred fir. In 2014 the team planted 360 *A. religiosa* seedlings gathered from ten different provenances in half a dozen plots in the core zone of the Monarch Butterfly Biosphere Reserve. The test site was placed at 11,286 feet elevation—deliberately a thousand feet or so higher than the current monarch butterfly roosting sites.

Oyamels are not responding well to the higher temperatures wrought by our changing climate. According to Sáenz Romero, the trees have reached their "xeric limit," the latitude and elevations at which they historically have been able to thrive. Sáenz Romero's higher altitude location choice proposes to address the reality of the inhospitable conditions the sacred firs are facing. The weather in Mexico has become hotter and drier—and the temperate mountains where oyamels have thrived for millennia can no longer support them. Sáenz Romero says that by 2090 not a single square kilometer of the Reserve will be climatically appropriate for *A. religiosa*. "The trees are going to die," he predicted with mild-mannered urgency. And with them, presumably, the monarch butterflies' long-standing roosting sites.

Sáenz Romero came to that conclusion after assembling an essay published in 2012.[2] He and U.S. Forest Service researcher Jerry Rehfeldt, now retired from the Forest Service, spent much of 2007–10 developing a model for how climate change affects Mexican forests and what can be done about it. After developing the model, the duo had to choose a single endemic Mexican species on which to extrapolate their projections. Rehfeldt suggested that since Sáenz Romero hailed from Michoacán, the winter home of monarch butterflies and the renowned sacred firs, *A. religiosa* was the obvious candidate. Sáenz Romero wanted to work on pines, but Rehfeldt said he convinced him to focus on the monarchs' favorite fir so as to "make a big splash."[3] Sáenz Romero has dedicated himself to the study and conservation of the storied tree ever since.

Native to central and southern Mexico, the oyamel casts a grand presence. A spiritual mood accompanies a walk among the impressive firs, which are used as Christmas trees and decorative foliage on church altars during the holidays. The Spanish name *oyamel* comes from the Nahuatl word *oyametl*.[4] The Latin *Abies religiosa*, given by German naturalist and explorer Alexander von Humboldt in the late eighteenth century, suggests the region's long-standing affection for the tree, which seems destined to become less enduring as the climate heats up.

Conical in shape and resembling hands folded in prayer, the tree so impressed Humboldt that he published the species description using the Latin word *religiosa* as its name. Spanish priests attempting to convert Indigenous peoples to Catholicism held mass in the forest under the vaulted cathedral-like canopy of the sacred firs because the local community refused to go inside a man-made structure to worship.[5] Like many of us, they preferred the church of the outdoors. As one recovering Catholic told me, that's why every church in Mexico has a courtyard or outdoor vestibule—because Native peoples refused to go inside.

In 2012, Sáenz Romero presented his article to scientists at the Monarch Butterfly Fund, the partner organization of Monarch Joint Venture in Minnesota. "They were kind of shocked—the trees are going to die . . . What?" he said, recalling the reaction to his presentation. At the time, a collective awareness of the impact of climate change on the forests that constitute the monarch butterflies' roosting sites did not exist. People in general and scientists who study monarch butterflies in particular were primarily focused on monarch butterfly biology, the impact of GMOs and pesticides, and deforestation in Mexico caused by illegal logging, mining, and farming. "It was a bit like a bombshell exploding," recalled Sáenz Romero.

The 2012 article added yet another wild card to the matrix of threats to the monarch butterfly migration. The presentation also earned Sáenz Romero a small three-year grant from the Monarch Butterfly Fund and a seat on the organization's advisory board. Funding from several Mexican government institutions and the Mexican arm of the Nature Conservancy followed.

Sáenz Romero's experimental forest reserve, Ejido La Mesa, sit on Sierra Campanario near San José del Rincón in the state of México. Formerly an ecotourism destination, La Mesa occupies only 0.22 hectares, about half an acre. In 1994 a fire at the colony forced the butterflies to relocate down the mountain to El Rosario, the largest and most visited of the sanctuaries. Tourists followed them, leaving the remote preserve and its tourist bungalows, signage, and sprays of blooming asters abandoned and underappreciated.

A visit to the La Mesa plot hints at what a monumental task the assisted migration of a forest represents. The 40- by 40-meter parcel, a ten-minute hike on a worn dirt path uphill from a rough, unpaved road, hosted three-year-old seedlings in various stages of success. Some looked like dried, brown nubs only four inches tall. More robust specimens flaunted limber evergreen shoots bouncing in the breeze and reaching two feet in height. An inspection of the plot recalled this gardening dictum: "The first year they sleep, the second year they creep, the third year they leap." How long would it take for the sacred firs to leap to the towering height and heat-generating girth that provide a protective blanket for monarch butterflies during their winter stay? Decades.

Our jaunty, uphill drive to La Mesa found Sáenz Romero pulling over repeatedly to provide context for understanding the forest. A vast green

Oyamel firs and other trees suffer from a lack of moisture in Michoacán, with clear-cutting visible in the background. Photo by Monika Maeckle

expanse interrupted with occasional patches of tan, cultivated soil unfurled before us. We jumped out of his Ford Tacoma under a warm spring sun. Sáenz Romero explained that the forest was recovering from erosion-induced landslides, the consequence of years of mining abuse.

What appeared to be a healthy forest to the amateur tree lover was, in fact, a forest pocked by illness and challenge. Sáenz Romero pointed to bare, naked treetops stripped of their foliage, randomly jutting from the vista. These oyamels suffered from dehydration, he said. Their scrawny tops and absent greenery suggested that they're not getting enough moisture.

Wet and dry seasons typify the weather pattern here in the Trans-Mexican Volcanic Belt. The mountain range straddles the states of Michoacán and México and is singular in running latitudinally, connecting the eastern and western legs of the Sierra Madre. It serves as a unique center of biodiversity in Mexico, hosting not only overwintering monarch butterflies but also hundreds of species of flowers, forbs, and endemic birds and mammals, in addition to the native conifers studied by Sáenz Romero. The region is important for its biodiversity and indigenous species, which probably were forced to migrate to the temperate mountain areas when the Central Mexican Plateau became drier and hotter during the Pleistocene era.[6] Historically, rain generally falls May through September here, and sometimes in October. The rest of the year is dry and temperate. Snow is not uncommon on the mountaintops.

When the wet season yields less rain than average and the dry season is hotter than usual, however, trees simply cannot draw enough water from the soil to sustain their highest branches. Transpiration, the loss of moisture from plant leaves, roots, and branches, increases. Leaf and branch shedding result, creating weak, skinny treetops. This compromised state also makes the forest more susceptible to insect attacks, wind damage, and dramatic weather, as demonstrated by a late-season storm in the spring of 2016.

On March 11 of that year, an unusual freeze and sleet storm descended on Mexico's Monarch Butterfly Biosphere Reserve. The tempest whipped through the area, clobbering more than one hundred acres of trees and killing an estimated 50 million butterflies. The disaster occurred at a most vulnerable time, since monarchs typically begin their journey north to lay their first generation of eggs in the first half of March.

In addition to the devastating tree and butterfly casualties, the storm left the international monarch butterfly community whipsawed. After a robust recovery in 2015 when the population tripled from 2013's historic low, the celebrated rebound of monarch butterfly numbers was stopped in its tracks. The 2016 storm pushed population counts back to 2014 levels.

The trauma from such storms lasts for decades. The March 2016 storm punched more holes in the heat-trapping forest canopy, which will haunt the roosting sites for years to come. Citing Lincoln Brower, Sáenz Romero reminded us that an intact forest canopy works like a blanket for the butterflies, insulating them when temperatures drop below freezing while they wait out the winter in their semi-hibernative state. That inactive state, induced by cold but not freezing temperatures, along with the protection provided by the forest help the butterflies conserve stored fats. Monarch butterflies reserve those lipids, built up during the fall migration when they nectar and fuel up on their way south, to power their spring return and reproduction. Holes in the forest blanket force the butterflies to burn through their fat reserves prematurely. And higher daytime temperatures in the forest drive them from their clusters on the sacred fir trees to the forest's shallow seeps and streams in search of water, burning more calories in the process.

Just as cactus won't thrive in the jungle or orchids in the desert, *A. religiosa* won't grow in inappropriate conditions. The tree exhibits a narrow range in which it can succeed: 9,200–11,500 feet elevation, with temperatures around 50 degrees and about 39 inches of rain per year. Given that volatile weather will continue as the mercury on local thermometers climbs, Sáenz Romero believes that a projected increase of 6.6 degrees Fahrenheit by the end of this century coupled with a dramatic 7 percent decrease in precipitation will decimate the monarchs' roosting forest. According to the models constructed by Sáenz Romero and Rehfeldt, by 2090 the oyamel trees that populate the mountains of the monarch butterflies' winter home will no longer be able to exist in their current locations.

Sáenz Romero's partner Blanco Garcia gazed at the expanse surrounding the La Mesa experiment and pointed out that the forest isn't recovering. It wasn't always this way, he said. "Sixty years ago, there was no forest here. . . . When the mining stopped in the area, the forest was able to regenerate itself." But that no longer appears possible given the assertiveness of a changing climate. "It's the speed of climate change," Sáenz Romero added. "If we had 10,000 years, it would be fine."

▼

The notion of deliberately moving plants out of their native habitat to a foreign environment challenges not only fundamental ideas of traditional forestry and land management but also our personal definitions of nature itself. In a 2011 article published in the *Forestry Chronicle*, authors Isabelle Aubin and C. M. Garbe laid out why the concept is so unsettling.[7] The greatest fear is that a translocated plant may adapt so well to its new environment that it becomes invasive, overtaking local indigenous species. Local species endemic to the plant's location may lose ecosystem benefits, resulting in a loss of biodiversity. Another possible outcome: failure at great expense and substantial risk.

These prospects loom in the minds of scientists and land managers. Our natural history is littered with examples of plants purposely or accidentally moved to places they don't naturally live, sometimes resulting in disaster. Take kudzu, *Pueraria montana*, for example. The aggressive Japanese vine arrived in the United States in 1876 when it was featured at the Philadelphia

Centennial Exposition. Billed as hardy, fast-growing, and attractive, with sweet-smelling purple flowers and sturdy structure, the vine was first presented to the gardening community as an ornamental. It ranked high in the "wow factor" that commercial growers seek.

In the 1930s, the Soil Conservation Service pushed kudzu as a solution for soil erosion. Once established through its aggressive runners and rhizomes, the vine can grow up to a foot a day. Dubbed "mile-a-minute vine" because of its hyperactive growth, kudzu kills native plants by hogging the light, smothering its neighbors under a thick leaf blanket. Its aggressive roots and shoots also girdle neighboring plants' stems, branches, even tree trunks by the sheer force of their weight. The unstoppably vigorous plant, once confined to the U.S. South, makes frequent appearances on lists of worst invasive species and marches north as our climate warms.

Kudzu is just one example. Chinaberry trees, feral hogs, Johnson grass, zebra mussels, Japanese honeysuckle, and the soil-poisoning *Eucalyptus* trees in California (ironically, a favored roosting site for West Coast monarchs) are some of thousands of invasive species that make assisted migration seem like a dangerous gamble for many. And yet the forest needs an assist from humanity, a growing group of scientists and activists believe. Assisted migration may be the answer, albeit controversial.

For the Torreya Guardians, a group of self-organized ecologists in Florida, the idea holds great promise. Chief tree evangelist and Torreya Guardians founder Connie Barlow considers her group "the radical edge of what is going to become a mainstream action."[8] The group has been working since 2009 to save the highly specialized *Torreya taxifolia* conifer from extinction by planting its seeds five hundred miles north of its traditional range. The evergreen conifer, known as "stinking-cedar" for the pungent smell its flesh exudes when burned, cut, or bruised, is historically found only in the high plateaus, steep bluffs, and deep ravines of the forty-mile stretch of the Apalachicola River in the Florida panhandle, not far from where Fred Urquhart suggested Lincoln Brower gather monarch butterfly specimens. The tree was listed in 1984 as endangered under the Endangered Species Act. At the time of its listing, all mature viable trees were located in botanical gardens and arboreta. As of this writing, *T. taxifolia* has lost 98.5 percent of its total population and has zero long-term prospects for survival.

Viewed as ecological terrorists by some and cutting-edge conservationists by others, the Torreya Guardians undertook the assisted migration of the Florida torreya by planting it 488 miles north in Waynesville, North Carolina. Members of the volunteer group took a cue from a specimen that was thriving at Biltmore Gardens in Asheville, North Carolina. The tree apparently was planted as a specimen there in 1939 by Biltmore Gardens botanist Chauncey Beadle.[9] As in Sáenz Romero's experimental forest, the survival of *T. taxifolia* appears to depend on moving it to higher ground with cooler temperatures.

Trees can't pick up and move like people, birds, or monarch butterflies. As Kara Rogers writes in *Quiet Extinction*, since no mountains exist in Florida, Florida torreya would have to relocate hundreds of miles north up the Chattahoochee River to survive. "It never made that trip on its own, of course," writes Rogers. "But we can make the journey for it, delivering it to the Appalachians to help it escape extinction. The question is whether or not we should." The Torreya Guardians did. And the trees, planted in various locations farther north, are thriving.

Assisted migration as a forestry and land management practice has been used elsewhere, especially in Canada. The endangered western larch, native to the northwestern Rocky Mountains, favors high altitudes, as does *A. religiosa*. The high-elevation species provides important high-fat pine nuts for grizzly bears, keeping the adventurous mammals from descending into towns in search of food. Hit by a lethal fungus, pesky mountain pine beetles, and a lack of forest fires that would keep such challenges in check, the endangered larch now must also contend with higher temperatures that projections suggest will kill 97 percent of the larch tree canopy by 2100.

In response, scientists planted the conifer in a remote area of British Columbia, five hundred miles north of its traditional range. A decade later, many of the transplants that are thriving came from a few hundred miles south—an indication of the dramatic march of climate change. The early data were so compelling that British Columbia's forestry agency officially adopted a policy in 2018 requiring foresters to use seeds from warmer-climate zones for the 280 million trees they plant each year.[10] "It restores the tree to the environment for which they are best suited," according to Greg O'Neill, an adaptation and climate change scientist with the government of British Columbia who helped design and implement the assisted migration program.[11]

Though moving a nonnative species into the wild disturbs many conservationists, scientists and land managers concede that something must be done. As mentioned in chapter 12, the USDA announced changes in plant hardiness zones and the map that communicates them in 2012, and again in 2023. Dozens of cities listed on the map were assigned new zoning designations.

The redefined hardiness zones tell us what butterflies, birds, and blooms have been communicating for years. The world is getting hotter. But unlike mobile species that fly, walk, or can be carried by wind to more friendly environs, trees can't move on their own. And they often take decades to mature and reproduce. As unorthodox as assisted migration may seem, the concept is gaining traction. Cities across the country managing their urban tree canopies for future resilience are routinely installing tree species that are native two hundred miles south of their prospective new locations. This will account for projected increases in temperature, climate change experts contend. By the time the trees mature, the climate will fit the tree.[12]

▼

Sáenz Romero's experimental forest trial continues. In 2023 he lamented the ever-lengthening dry season, which he said now runs from November through May. "It's putting a lot of stress on our seedlings," he said, "not massive mortality, but they're not in good shape."[13] The trees need partial shade during the harsh dry season to thrive, so Sáenz Romero and his team have tweaked their approach in recent years and planted oyamel seedlings in circles, interspersed with *Baccharis conferta*, known as *escoba* or broom plant. A member of the aster family native to the Trans-Mexican Volcanic Belt, broom plant can reach six feet in height.

The plant palette surrounding the oyamel seedlings matters, Sáenz Romero explains. Sacred fir trees planted in biodiverse combinations with sage bushes and tall perennials like *B. conferta* that provide shade and help retain soil moisture fare better than those placed unprotected in full sun. The trees can also benefit from a diverse plant community that fosters their symbiotic relationships with mycorrhizal fungi, which typically colonize plant roots and trade nutrients with them in a network that has become known as the

The seedling of oyamel fir in the center died because of a
lack of water. Photo by Monika Maeckle.

"wood-wide web" in recent years.[14] The more traditional approach to refor-
estation, clear-cutting, just won't work in this instance, Sáenz Romero says,
nor will planting trees in traditional forestry's tidy rows.

Even if Sáenz Romero's experimental forest thrives and the monarchs'
chosen roosting tree, *A. religiosa*, succeeds in its experimental plot, the battle
to save the tree will continue. By the turn of the century, even a roosting site
a thousand feet higher up the mountain will not be able to endure the higher
temperatures predicted to result from our changing climate. The experimental forest at La Mesa is a makeshift solution, simply buying the monarch
butterflies and the sacred firs some time. "Assisted migration is a way to
mitigate the situation, but it's not a perfect solution and a perfect solution
does not exist," advises Sáenz Romero, adding that the forest will need to be
moved yet again, perhaps to Popocatépetl, Mexico's highest mountain and
second-most-active volcano, which sits at 5,452 meters, almost 18,000 feet.
That's about five thousand feet higher than the current tree line—the altitude
at which trees have historically been able to grow—in this part of Mexico.

Monarchs are already moving up the mountains. In February 2018, the
Mexican media company Milenio documented a roost at the foot of Popocatépetl about seventy kilometers (forty-three miles) southeast of Mexico City.
According to the four-minute video, a small colony of monarch butterflies has
been taking refuge here for years in a ravine southeast of the volcano known

as Joya Redonda ("Round Jewel"). Monarch scientists say they have known about the roost for years, but it hasn't been studied because it's outside the Monarch Butterfly Biosphere Reserve and extremely inaccessible.

The area is controlled, once again, by a collective ejido, so the government cannot dictate how it is managed. An informal troupe of environmentalists conduct tours and try to instill respect and order there, but visitors "grab the butterflies, they step on them, they want to take them, they go in with food," according to one volunteer guide.[15] The group wants the Mexican government to intervene on behalf of the butterflies.

In early 2019, World Wildlife Fund officials announced that, for the first time, a monarch roost was discovered at Nevado de Toluca in the state of México. The stratovolcano, a conical volcanic landform resulting from layers of lava and ash building up from successive eruptions, sits fifty miles west of Mexico City at 4,680 meters, or 15,350 feet elevation, the fourth-highest peak in Mexico. Like our sea levels and temperatures, monarchs appear to be adapting to our changing climate by moving ever higher.

▼   ▼   ▼

CHAPTER 16

CAPTIVE BREEDING, POLLINATOR DRONES, AND MONARCH ZONES

The assisted migration strategy of relocating trees grown in greenhouses to save the forest where the monarchs roost has become more acceptable to the science community in recent years, yet releasing monarch butterflies bred in captivity at the roosting sites in the event of a catastrophic weather event has remained a nonstarter. Why?

The idea, mentioned by IBBA president David Bohlken (see chapter 14), surfaced at the June 2018 gathering of the CEC. Bohlken accompanied wife Jane Breckinridge, who was invited to the meeting as an ambassador of Oklahoma's Native American community. Bohlken and Breckinridge own and operate the Euchee Butterfly Farm in Bixby, Oklahoma, about two hours west of the state capital. The farm sits on a 160-acre property that was deeded to Breckinridge's great-grandmother, Neosho Parthena Brown, in 1899. Brown was the daughter of the last Euchee chief, Samuel W. Brown.

The CEC meeting celebrated the organization's twenty-fifth anniversary that year with a focus on innovation and partnership, according to an EPA press release. EPA administrator Scott Pruitt, chair of the CEC at the time, led a visit to the National Weather Center in Norman, Oklahoma. Pruitt announced two new committees that would tackle an "extreme events mapping initiative" and "extreme events advisory group," carefully avoiding the words "climate change," as per then-president Donald Trump's discouragement of such verbiage. The charge of these new committees: preparedness and response in the wake of "extreme events," with a focus on drought, wildfires, and extreme temperatures.[1]

According to Bohlken, representatives from Mexico approached him during the meeting to discuss the possibility of developing a plan for commercial butterfly breeders to "help save the migration" by producing butterfly livestock for release at the Monarch Butterfly Biosphere Reserve. Specifically, the representatives were curious about the possibility of a captive breeding program at the preserves, and the potential for restoring populations in the event of a catastrophic weather event similar to 2016, the year a freak sleet storm in March decimated 30–38 percent of the overwintering population.[2]

The likelihood of such an event happening again is high, the group agreed. And with a migratory monarch population that vacillates wildly depending on weather each year, concerns that an extreme weather event could take out the roosting population, especially in a year when the numbers were down, were justified.

As chief representative of the premier trade association for the commercial butterfly breeding community worldwide, Bohlken pounced on the idea. He reached out to the organization's top breeders and identified several IBBA members willing to assist. The plan would require $100,000–$125,000 to buy fresh tropical milkweed seed every three years, the group determined, because milkweed seed is highly perishable. "It would be a Mexico-led effort," said Bohlken, adding that a third of the butterflies would be bred in Canada, a third in the United States, and a third in Mexico, to replicate in some fashion the provenance of the migrating population.

An economist by training, Bohlken had already run the numbers and identified funding sources. Two granting organizations had committed to underwrite the effort. Bohlken had already identified seed sources in Surinam for the controversial *Asclepias curassavica*, tropical milkweed, on which to grow the insects. After securing seed, Bohlken said, growers would have to move fast since *A. curassavica* seed loses viability quickly. Also, the plan would work only if the catastrophic weather event occurred in early February, since it would take six weeks to get the plants growing.

Could such a scheme actually work? Monarch butterfly scientists say no, but programs to rebuild endangered and threatened species populations—including butterflies—through captive breeding have proven successful elsewhere. Gray wolves were brought from Canada to Yellowstone Park in Montana in the 1990s to restore the decimated population there. Peregrine falcons also rebounded after coordinated efforts to transfer at-risk chicks from one area, install them in nest boxes elsewhere, and foster them temporarily with limited human contact, a process called "hacking." The birds are now seen regularly in large cities and coastal areas.

Perhaps the most relevant model is the *Phengaris arion*, commonly known as the large blue butterfly, which disappeared from England in 1979. In the course of ten years, scientists brought it back. Jeremy Thomas, professor of ecology at Oxford University, determined the unusual life cycle and specific needs of the pale blue butterfly: wild thyme flowers and a particular species of red ant. The large blue caterpillar consumes the thyme flower head before turning into a carnivorous caterpillar. In its early stages, the caterpillar eats its competing larvae, but later it drops to the ground, hoping to be discovered by a red ant. When an ant taps the larva, it secretes a sugary liquid that the ant devours, and the two creatures carry on this mutually beneficial relationship for up to four hours. At some point, the large blue contorts its body in such a way that the ant mistakes it for an ant grub and carries it underground to the nest, where it will spend months with red ant grubs. There it forms a chrysalis. Upon hatching, the large blue uses its wings to tunnel to the surface to take flight.

David Simcox, a conservation consultant for the Centre for Ecology and Hydrology, retrieved large blue eggs from a population on an island in Sweden. He fostered the eggs in captivity on thyme leaves and released the caterpillars in Devon and Somerset, England, where the particular red ant species thrives on managed pasturelands.[3] By 2018 the large blue was recorded at forty sites in Great Britain, marking its best year on record. Thomas cited the large blue as the first butterfly in the United Kingdom to replicate population numbers seen previously at its peak. "Our approach has now become the model for insect conservation worldwide," Thomas told the British publication the *Guardian*.[4]

Monarch butterfly scientists remain skeptical of the approach discussed by Mexican officials and Bohlken. Upon hearing the idea, Chip Taylor of Monarch Watch, with whom Bohlken had worked on a program to train Native peoples in Oklahoma in best butterfly breeding practices, responded, "It can't be done."

Bohlken shared the idea with Wayne Wehling, a USDA entomologist who for years oversaw the permitting of butterflies shipped across state lines within this country as well as exotic species entering from outside the country at the time. Wehling, who has since retired, assured Bohlken the USDA would be on board with appropriate permits required and the "government acumen needed to do things" for a trilateral captive breeding effort to save the monarch migration.[5]

But Wehling also admitted to having reservations. "I've got to scratch my head on the scientific or environmental considerations as to whether that's worthwhile or not," Wehling said, adding that it takes some "seriously hard natural selection" for butterflies to make the migration to and from Mexico.

Migration expert Andy Davis, who happens to be married to monarch butterfly disease specialist Sonia Altizer, labeled the idea of releasing captive reared butterflies at the monarch roosting sites in Mexico "a little too far-fetched." "It sounds misinformed. It sounds like it's based on no science at all." Davis wondered aloud whether those who proposed the possibility had any inkling of monarch biology and life cycle.[6]

"This would be an experiment with no predictable outcome and difficult to assess if it were successful," said Carlos Galindo Leal, who served as director of science communication for CONABIO, Mexico's National Commission for the Knowledge and Use of Biodiversity, at the time. He suggested investing in habitat as a better use of resources.

Taylor, Anurag Agrawal, and Karen Oberhauser all expressed exasperation and disbelief that the idea would even be floated. "We can never breed our way out of the monarch crisis," said Oberhauser, who added that such an initiative would take the focus away from solving the broader conservation challenges that caused the decline.

▼

An oversized honey bee lights on a perfect Gerber daisy on a clear sunny day, somewhere in Great Britain. The bee looks a little odd—bigger than average, its wings completely transparent, body idealized, pristine. The buzzing sound it makes has a mechanical pitch, like the twitter of a small chain. The bee shakes its backside on the flower as a female newscaster announces in a British accent, "The honey bee–mimicking drone insects known as ADIs have been activated for their second summer."

So begins "Hated in the Nation," one of the creepiest and most memorable episodes of *Black Mirror*, a science fiction anthology that offers a twisted take on the pervasive and invasive technologies to which we are addicted. Called a "Twilight Zone for the digital age," the series explores how the reflective dark screens into which we peer much of each day impact our nature and humanity. In this especially dark episode, someone hacks the computer system of the Granular Project, a large British company that has responded to the extinction of bees by creating armies of remote-controlled pollinators. These robot bees, known as Automatic Drone Insects, or ADIs, have made Great Britain's food and drink possible in recent years, but apparently they can also be programmed to kill people.

You'll have to watch the episode to see what happens, but be advised that, although no man-made technology can match bees for executing pollination chores, that hasn't kept companies of all stripes from trying. In March 2018, Walmart announced its intent to patent "systems and methods for pollinating crops via unmanned vehicles"—in other words, pollinator drones.[7] The eighteen-page application laid out a plan for the world's largest grocer to continue expanding its 23 percent market share in supplying food to households around the world.[8] The patent application suggested that one or more unmanned aerial vehicles (UAVs, commonly known as drones) would include a pollen applicator configured to collect the fertile powdery substance from one crop and apply it to another. A sensor would then follow to verify that pollination had been successful. Data would be reported back through an onsite network to an electronic database.

Within six months of Walmart's application, Dutch scientists of the DelFly project at Delft University of Technology announced their success at replicating the flight movements of fruit flies. The researchers were intrigued by these insects' ability to maneuver and avoid being swatted. Rather than use helicopter-like propellers like most drones do, the DelFly team replicated fruit fly aerodynamics, resulting in a tiny drone that can fly at 15 miles per hour, beat its wings seventeen times per second, and twist, turn, and pivot like nothing before. "The use we see for this is pollination in greenhouses," researcher Matěj Karásek told the *Guardian*.[9] The bee is under threat due to our farming methods, and we don't know what their future will be. This is one solution."

But as with previous agriculture "systems," red flags waved high and mighty. The Xerces Society came down hard and fast against pollinator drones. In the spring 2018 edition of *Wings*, the invertebrate conservation organization's quarterly magazine, Scott Hoffman Black and Eric Lee-Mäder acknowledged the research occurring at some of the most distinguished universities in the world but reminded us that it's not that simple. Crop plants sport myriad flower shapes and sizes, and, generally, a custom insect pollinator has coevolved with these plants over millennia. A one-size-fits-all pollinator drone just won't cut it, the Xerces team pointed out.[10]

And it's not just about crops. About 85 percent of all terrestrial plant species benefit from a variety of insect pollinators. The honey bee, though a spectacularly efficient pollinator, represents just one of more than a thousand bee species. It has a particular shape, size, and set of habits. Bumblebees and solitary bees, for example, have the capacity for buzz pollination, a specific type of pollination required by about 9 percent of flowering plants in the world. Buzz pollination requires a larger-bodied insect to vibrate its flight muscles rapidly to dislodge sticky pollen from a plant's anthers. Tomatoes, peppers, and some berries require buzz pollination—thus industrial growers are tapping bumblebees for their pollination services inside greenhouses.

Given the coevolution of bees and other insects with the plants they pollinate, how many different types of pollinator drones would a farmer need to deploy to fertilize his or her crops efficiently? And what would that cost?

David Goulson, professor of biology at the University of Sussex and founder of the Bumblebee Conservation Trust, posed those questions not long after the Walmart story broke. While acknowledging the intellectual interest in developing robotic bees, Goulson dismissed the notion as "remarkable hubris."[11] Consider the numbers: roughly 80 million honey bee hives populate our world. Each hive contains about 40,000 bees through spring and summer—a total of 3.2 trillion bees. "They feed themselves for free, breed for free, and even give us honey as a bonus. What would the cost be of replacing them with robots?" Goulson asked. And this doesn't include the other 20,000 species of bees that contribute to pollination chores.

He went on to suggest that even if the robo-bees and their associated power packs and control devices could be produced for a penny apiece, costs would exceed $32 billion. Add to that the costs of staffing, electricity, technical malfunctions, weather issues, wildlife intervention, and cleanup of downed drones. "What about the environmental costs of manufacture? What resources would they require, what carbon footprint would they have? Real bees avoid all of these issues; they are self-replicating, self-powering, and essentially carbon neutral," said Goulson.

This technological drama will continue to unfold as the world's irreplaceable band of free ecosystem service workers continues its demise and several species of bees join the growing list of endangered species.

▼

As *Homo sapiens* pushes the boundaries of combatting pollinator decline, a troupe of monarch butterfly advocates in the American Midwest has taken on a more hands-on, low-tech approach. At a former thoroughbred horse-breeding facility known as Clearwater Farm in Linn County, Iowa, in 2018, hundreds of fourth and fifth instar monarch caterpillars replaced prize-winning show horses as the preferred livestock. The hungry striped larvae noshed on common milkweed in a mesh enclosure inside the barn's lab area. In the back, dressage halters and fancy saddles once occupied the storage area, but now shovels and hoes, hundreds of plant pots, and stacks of boxes filled the stable's tidy stalls. Known for the clear blue waters of its man-made ponds, the property houses the Monarch Research Project (MRP), a small nonprofit organization driven to help save the monarch butterfly migration.

The MRP's signature program is Monarch Zones, an initiative that crowd-sources the rearing of monarch butterflies to the local community by providing everything needed to raise them in trademarked BioTents. The fostered livestock is released to the skies, with a hope of increasing the number of migrating monarch butterflies. In its first three years, MRP claimed to have raised and released about 55,000 monarch butterflies.

Participants purchase special rearing enclosures from Monarch Zones. Becoming a Monarch Zone participant, which costs $230–$417, includes a BioTent enclosure, access to milkweed plugs, monarch butterfly eggs, caterpillars, even fecund females for laying future eggs. The MRP also provides a user manual—thirty-nine pages of instructions on how to raise monarch butterflies. "As a Monarch Zone you will raise monarch stock for 30 days, releasing them after completion of MZ protocols," reads the welcome card and member agreement that accompanies Monarch Zone membership.

Michael Martin, research station manager who oversaw Monarch Zone's day-to-day operations until his untimely death in 2022, explained what he called a "one and done" approach to "rearing monarch butterflies on the wild side" during a tour of the facility. First, identify a suitable location for your Monarch Zone. Cover the space with landscape cloth that comes with your kit—the precut holes in its fabric allow for easy planting. Install milkweed plugs in the holes. Then place the BioTent over the milkweed-planted landscape cloth, add monarch butterfly livestock to the enclosure, zip up the door, and walk away.[12]

"You can close it up and never have to go in again until you see butterflies flying," said Martin, a master gardener and professional landscaper who worked for years at nearby Kirkwood Community College. The specially selected mesh allows rain or supplemental water into the enclosure without even unzipping the door. Ample airflow keeps the insects healthy, and the enclosure keeps out predators like birds, wasps, tachinid flies, and possibly diseased wild butterflies.

Monarch Zones began in 2014 at the former thoroughbred horse farm owned by Clark and Marian McLeod in Marion, near Cedar Rapids. The eighty-acre spread sits in the heart of the monarch butterfly summer breeding grounds. Clark McLeod was taking a lesson from his tennis coach, Cam Watts, who introduced him to the magic of monarch butterflies. Watts, a former science teacher, had been raising the insects for years. After McLeod learned the story of the monarch migration and its decline, he decided to apply his energies and resources to the challenge of rebuilding the butterfly population. "The monarch is magical, it pulls you into the conundrum," said McLeod.

Perhaps best known for his career as a self-made billionaire telecom executive whose company McLeodUSA became the subject of multiple lawsuits in the mid-2000s when it went bankrupt, McLeod likes to apply business principles to what he calls "problems that people are dancing around." Monarch Zones became the pilot for what he envisioned as a three-pronged national program to help save the monarch migration. He described the MRP's goals as "moonshots," a term typically used to describe bold, exploratory or groundbreaking approaches that often include an incomplete understanding of potential risks and benefits.

Moonshot 1, McLeod explained, was to replace the monarch population in Linn County by creating Monarch Zones; he preferred to focus on a specific geographic area during this pilot period. Moonshot 2 aimed to create 10,000 acres of habitat in Linn County by partnering with cities and counties, private landowners, and public land managers. And Moonshot 3 hoped to qualify and quantify the success of the project into a "playbook" to be shared nationwide for "Zones across America."

But as with Sáenz Romero's experimental forest in the Mexican mountains, those involved in the Monarch Zones have their detractors. Chemical ecologist and monarch and milkweed researcher Anurag Agrawal is bothered by "the audacity" of the approach and compared the systematic introduction of captive bred monarchs into the wild to "ladling water out of a sinking ship." "It will make some of us feel good, and we might learn some butterfly biology along the way," said Agrawal, which are good but insufficient justifications. Such approaches, he argues, contribute to "plausible deniability about our environmental problems as well as our superiority complex as a species." Sonia Altizer, one of the foremost experts on monarch diseases, especially OE, expressed the usual concerns of raising lots of monarchs in close quarters. "How do we know the monarchs are healthy and disease free?" she asked. Even small lapses in screening (missing one infected monarch) can lead to a disease outbreak, and not just OE. "Every few years in my lab we have catastrophic larval and pupal mortality due to non-OE causes—and that's even when rearing monarchs singly with clean plants and sterile containers."

Martin contends that MRP abides by best practices developed by commercial butterfly breeders, with whom MRP consulted on their protocols. He and Watts attended workshops staged by the IBBA and Association for Butterflies to learn procedures and protocols. Breeding stock was tested for OE, and milkweed leaves were triple-washed in a 5 percent bleach solution and dried in a salad spinner before feeding to caterpillars. A tour of the facility in 2018 required stepping into a tub of bleach solution before entering the caterpillar-rearing area as a safeguard. In addition, MRP also limited the number of eggs and caterpillars supplied per BioTent to thirty-four—typically one per milkweed plant. "We test for OE and have only had one case," Martin said. "For us, it's about rearing right along with nature, providing as natural an environment as you can get outside."

The Monarch Zones experiment continues. In 2019 the organization began offering a revised version of its original BioTent program, one in which the grow houses are placed over existing patches of common milkweed, *Asclepias syriaca*, the large-leaved milkweed commonly found in the Midwest. This was an effort to take advantage of existing prairie situations. "We feel like we're doing some good here," said Martin. "We're making a difference. If we don't do something, then who? Coulda, shoulda, woulda—we don't want to be caught saying that."

CHAPTER 17

▼ ▼ ▼

CHANGING RHYTHMS
OF THE PLANET

I might be walking my dog Cacteye along the San Antonio River in early spring when it happens. Or maybe I'm paddling the Llano at the ranch as the sweet honey fragrance of agarita blossoms wafts in the breeze. And then I'll see it: my first monarch butterfly of the season.

In San Antonio, first-of-season monarch sightings are usually in late February or early March. In the Monarchy, these moments are often noted as an "FOS" sighting, an acronym also used in the bird world, along with "FOY," for first of year. Posts on the D-PLEX list and social media hail these occasions, frequently accompanied by effusive language, exclamation points, emojis, and photos. Typical posts: "Saw my FOS monarch today near Point Pelee." "FOS monarch eggs in central Arkansas!" "First-spotted monarch 😊 instars of 2020!" "Video of first of season monarch seen here at our waystation. . . . YAY!"

FOS alerts are not limited to butterflies. They can announce eggs, caterpillars, chrysalises, roosts, or milkweeds. We seem to crave these seasonal moments, awaiting them with anticipation. And when they occur, signaling a new beginning, they coax a smile and a fleeting assurance that the rhythms of the planet are intact and "everything's gonna be okay."

FOS tracking through citizen science projects and social media is common practice. Journey North, Correo Real, Monarch Watch (see chapter 5), and more recently the crowdsourced nature observation app iNaturalist all provide accessible updates and data points of where and when monarchs have arrived, as well as where milkweed and wildflowers are observed.[1] Journey North correspondents in Mexico share weekly updates from the roosting sites. As a new season revs up and the butterflies become activated,

"Monarchs are on the move!" and "Monarchs heading your way!" become common refrains.

The timing of a monarch butterfly's departure from Mexico and its arrival north is largely determined by day length, temperature, and climate conditions—the science of seasonal change known as phenology. Bird expert and writer Scott Weidensaul calls phenology "a $10 word for the timing of the seasons."[2] Others consider it climatology with a major biological component. Phenology is really a mash-up of all those.

Climatology studies the data gathered by instruments—air pressure, wind speed, precipitation, humidity, and temperature. Phenology is more complex because it considers not only those measured variables but observations of their biological impact on natural events and systems. The USA National Phenology Network defines phenology as "nature's calendar." The changing seasons are marked cyclically by the sprouting of leaves, the budding of flowers, the release of pollen, the formation of fruit, the dropping of leaves—as well as the appearance of monarch butterflies and other creatures at specific times and latitudes.

In nature, timing is everything, and the biological definition of success is the perpetuation of the species. If a plant sprouts early in the spring as a response to unseasonably warm weather, and then a freeze hits, the plant may be doomed. Depending on the plant's stage of development, it may be frozen in its tracks and consequently lose its chance to flower, fruit, and reproduce—a biological failure. Other life forms dependent on that plant—for food, shelter, or as a host plant—may also suffer when predictable seasonal rhythms are disrupted. These natural synchronicities are relied upon by all forms of life and can result in success or failure for plants and animals.

In recent years, these natural rhythms have become increasingly rattled and are causing what scientists call a "phenological mismatch"—that is, a system that's out of sync. When a previously synchronized biological event (like the emergence of milkweed) fails to occur in time to meet the needs of another biological event (like the arrival of fertile female monarch butterflies), disruption and damage can result.

Storm Uri, the historic Texas ice and snowstorm of February 2021 that resulted in 151 continuous hours of freezing temperatures, serves as an example. The "Snow-Mageddon" killed hundreds of people and millions of animals, knocked out electricity for half the state, and resulted in an estimated $295 billion in economic damage.[3] Just weeks prior to the storm, temperatures hovered around 80 degrees in San Antonio, the heart of the Texas Funnel. Then on February 15 temperatures dropped to 9 degrees and stayed in the teens for several days.

Birds, bats, fish, deer, plants, and other life were killed by the extended freeze. Many native plants were stalled but survived, perhaps forgoing reproduction for a season. Warm temperatures weeks earlier encouraged native wildflowers and milkweeds to push from the earth in late January. They were all set to flourish, flower, and eventually seed. But just as they were getting ready to do so, the storm hit, leaving dead buds in its wake. Native milkweeds in the Texas Hill Country stalled in the frigid soil. Tropical milkweed along the San Antonio River froze solid. Wildflowers were delayed. The result: not much nectar and not many milkweeds greeted monarchs upon their arrival.

According to reports from the community science organization Journey North, monarchs left their Mexican roosting sites earlier in 2021 than ever because of cues from the warmer-than-average temperatures. By March 5, observers on the ground in Mexico were reporting that monarchs were flying and copulating, but the clusters that form what is considered a roost or colony had dispersed.[4]

Monarch butterfly FOS sightings followed, and they were reported to be largely concentrated in Texas, with the migration's leading edge stretching into Oklahoma and Arkansas by early March. The butterflies had to continue flying because the milkweed and nectar that typically await them had been delayed and, in some cases, canceled.

The success of the monarch migration depends on a predictable schedule of seasonal events. Cold weather in the winter. Rain. A gradual warming of the earth. The early growth and eventual flowering of milkweeds and nectar plants. Lack of predictability based on seasonal cycles paves the road for a failed migration because, when female monarchs arrive from Mexico and find neither nectar for fueling nor milkweeds for reproducing, they'll continue flying north where those resources are even less likely to be available. The chance of them dying before reproducing increases. And since biology defines success as reproduction, those monarchs, and the migration they undertook, will have failed.

How a changing climate and phenology will affect the future of monarch butterflies and their migration is unclear. The insects traverse regions of several different climates, all of which are adapting in different ways to a warming world. Undeniably, winters are getting shorter and warmer. Spring is arriving earlier and quickly morphing into hotter, drier summers that are lasting longer. And the range for the oyamel firs where the monarchs roost—like our temperatures—is moving ever higher.

As the arrival of spring comes earlier, so do the consequent biological events, like the emergence of milkweeds and nectar plants, both of which monarchs need to survive. These uncoordinated natural events beg the question: can monarchs and milkweeds adapt?

They already are. Monarchs are departing the roosts weeks earlier than historically recorded. They're moving north ahead of their traditional schedule. Eggs are being spotted earlier in the year on young milkweeds, and barring a late season freeze, caterpillars are hatching and forming chrysalises that become butterflies. The life cycle continues. In some years, extra generations of monarch butterflies result from the early start and increased availability of milkweed.

In response to Storm Uri, Monarch Watch joined forces with iNaturalist to perform a census on monarchs, nectar sources, and milkweeds to determine which plants monarch butterflies were accessing in the wake of the historic freeze. More than 27,000 observations of 1,183 species by 4,670 observers were collected in the weeks following the storm. The project found that monarchs visited sixty-nine plant species in the spring of the historic Texas freeze. These species included native and nonnative plants. Though this number was fewer than the eighty-five species observed in 2020 and involved plants seldom used in good years, monarchs appear to have adapted to the new conditions.

Chip Taylor summarized the results on the Monarch Watch blog. Though acknowledging the study's limitations, including its humancentric concentration around major cities that are most accessible to volunteers, Taylor suggested that the findings show monarchs are "extremely adaptable" and can shift to new flower sources when needed by tapping a variety of nectar sources beyond those that appear naturally. In short, monarchs are opportunistic feeders completely capable of learning, and though the insects likely

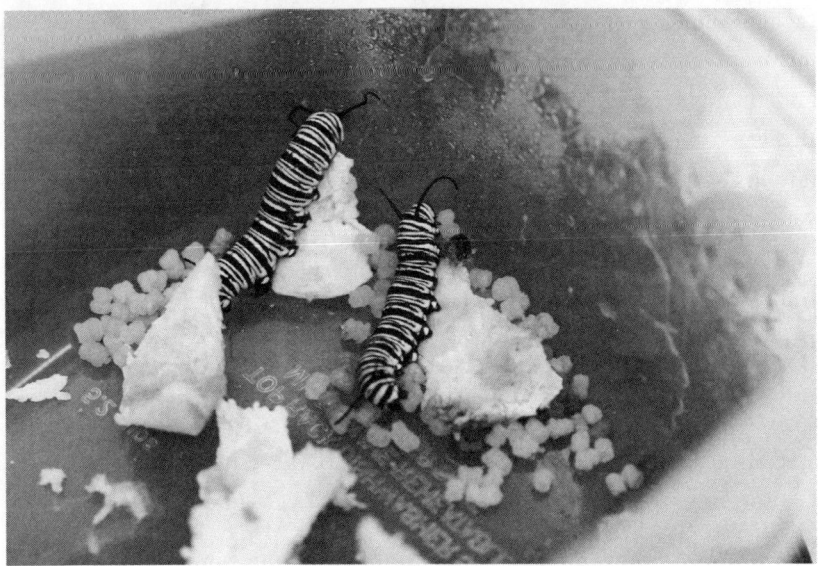

Monarch butterfly caterpillars eating pumpkin. Photo by Edith Smith

prefer high-quality nectar plants in good years they'll eat nearly anything when resources are scarce.[5]

Another food source monarchs have experimented with when faced with a lack of milkweed: pumpkin. In another example of the adaptability of *Danaus plexippus*, hungry monarch caterpillars have been known to consume pumpkin, squash, and cucumber in their late stages. Members of the Monarchy who raise monarchs at home as well as professional butterfly breeders—even scientists—have provided such alternative fuels when they've run out of milkweed or have been conducting experiments.[6] Butterfly breeder Edith Smith experimented with various members of Cucurbitaceae, the gourd family, at her farm and found that monarch caterpillars survived the menu change only if the alternative foods were introduced in the fifth instar. Any earlier, and the caterpillars were stunted and died.

Upon hearing of the alternative foods for monarchs, David James, entomologist and monarch researcher at Washington State University, conducted an informal experiment in 2020. He tested early monarch instars to see which foods the monarchs would accept and feed on—not just members of the gourd family but potatoes, tomatoes, apples, pears, and blueberries. The caterpillars ate everything offered for several days before perishing, except for the potatoes and blueberries, which they refused. The pumpkin, squash, and cucumbers were the most popular choices. And, interestingly, Lincoln Brower somehow convinced monarch caterpillars to eat cabbage for his famous barfing blue jay photo.[7]

Chemical ecologist Anurag Agrawal speculated that the chemical similarities of tetracyclic triterpenes in cucumber plants and the cardenolides in milkweeds represent a possible feeding link but conceded that no one has really studied this phenomenon.[8]

In another show of resilience and adaptability, the same year that the eastern population of monarchs navigated Storm Uri, monarchs in California staged a stunning recovery. In November 2020, western monarchs counts in California showed only 1,914 butterflies at 246 overwintering sites, an average of 7.8 butterflies per site. In 1997, the year the counts began, an average of 12,233 butterflies had populated each of the 101 sites, totaling more than 1.2 million butterflies.

Headlines and nonprofits predicted the extinction of the California population. A grassroots coalition formed after the dramatic decline. California launched a $1 million state-funded initiative to restore the western monarchs' natural habitat by planting six hundred acres of milkweed statewide.[9] And in February 2021 the California Department of Fish and Wildlife made it illegal to touch, net, rear, or handle monarch butterflies in any way without a special permit. The mandate became known as the "hands-off rule."

A year later, California monarchs had pulled off their impressive rebound. Following the record drop in their numbers to fewer than two thousand in 2020, the butterfly population vaulted to 247,237 in late 2021. The Xerces Society, which organizes the annual counts, called the recovery "remarkable." That this dramatic increase occurred in a year when California endured historic wildfires, a late season "bomb cyclone" rain event, and record air pollution is equally noteworthy.

Scientists were stunned, and slightly relieved. As Monarch Joint Venture noted at the time, the good news provoked excitement, relief, and celebration as well as substantial uncertainty about how such a positive development came about and what it suggests for the future. David James, who focuses his studies on the western monarch population, has some theories about that. He asserts that a warming climate and the availability of nonnative milkweeds in the winter, especially in urban heat islands and gardens around San Francisco and Los Angeles, played a sizable role in the 2021 monarch recovery.[10] Gardeners are increasingly planting tropical milkweed, *Asclepias curassavica*, and African balloon plant, *Gomphocarpus physocarpus*, in their yards and landscapes. Unlike native milkweeds, these species often do not die back in the winter, and they thrive in warmer climates. The presence of these year-round host plants on which monarch butterflies can lay eggs affords the butterflies new opportunities to reproduce and grow their populations throughout the year, thus boosting population counts. One example: the rise of the red admiral butterfly in 2023.

The red admiral, *Vanessa atalanta*, saw a 400 percent increase in its reported population in the summer of 2023—thanks to climate change, according to Butterfly Conservation, a nonprofit conservation organization based in Britain. An article in the *Guardian* touted the news: "Red Admiral Butterfly Population Soars 400% in UK as Winters Warm," read the headline, with the story explaining that, though the distinctive orange, black, and white brush-footed butterflies typically have historically traveled to the United Kingdom from North Africa and continental Europe, "in recent years the species has been overwintering in the UK rather than flying to warmer countries, as the weather is now not too cold for it to survive."[11] In the past, the butterflies would head to Britain to lay eggs in spring, then fly back south. But as a spokesperson for Butterfly Conservation told reporter Helena Horton, "with the increased frequency of warm weather, the UK may well become a permanent home for this species."

James has seen this movie before. His monarch studies began in the 1970s in Sydney, Australia. There, where monarchs were introduced in the 1850s, probably via cargo ship or from an island-hopping migration, he documented winter breeding monarchs on nonnative milkweeds right next to large clusters of nonbreeding monarchs in nearby trees. The juxtaposition challenged the conventional monarch wisdom that reproductive and nonreproductive monarchs do not coexist side by side. When the historic California rebound occurred, James felt a sense of deja vu, declaring himself "probably the only unsurprised monarch scientist."[12]

James believes that monarchs are extremely flexible, which is why you see more reproduction in the American South and California than witnessed a decade ago. The small overwintering count of monarchs does not represent the total California population. Thanks to a rapidly warming climate and wider availability of host plants, monarchs are reproducing in what previously was the season in which they went into a state of reproductive diapause. They're spreading out and expanding their breeding range, which makes them more difficult to count.

With winter breeding by monarchs on the rise and nonbreeding populations on the decline, California in 2021 recalled for James his monarch studies in Australia decades earlier. Overwintering colonies in Sydney in the early 1960s dropped from 20,000–40,000 monarchs per site to 1,000–3,000 per site in the late 1970s—more than a 90 percent decline. Regular formal counts of monarchs in Australia do not occur as they do in the United States, since both monarchs and milkweeds are nonnative species there.

James returned to Sydney in 2019 to revisit the locations of his former studies and research the monarchs' population status. He found no further radical decline in monarch numbers since the late 1970s.[13] Interestingly, Sydney sits at 33.51 degrees latitude south of the equator—about the same latitudinal distance from the equator as the California roosting sites in San Diego.

▼ ▼ ▼

CHAPTER 18

ENDANGERED OR NOT?

The question had been hounding me for years—ever since Lincoln Brower had joined with three nonprofit organizations to petition the U.S. Department of Interior to list monarch butterflies as "threatened" under the Endangered Species Act. He and I were chatting on the phone some time in 2016 when I finally blurted out the question: "Do you *really* think the monarch is threatened by extinction?"

Brower paused for a long moment. I thought of the hundreds of email exchanges and phone chats I'd had with him in recent years, conversations in which he had always been forthright. "Don't you think it's done some good?" he responded.

Though he never answered, it was clear that Lincoln Brower, the legendary monarch researcher who studied the insects longer than anyone else in history, knew his favorite insect would not be going extinct. Brower often referred to the "extraordinary biological phenomenon" represented by their multigenerational migration and was quoted saying that we should care about monarchs for the same reason we care about the *Mona Lisa* or Mozart's music. He considered the migration a treasure. Many professional scientists, citizen scientists, and others have expressed their devotion of so much time, energy, and money saving the monarch migration as motivated by a desire to "show the grandchildren" this unique marvel of nature.

But since no legal or regulatory mechanism exists for protecting biological phenomena, Brower and others supported a false narrative that would inspire hearts and minds: the monarch butterfly is going extinct. Brower was a pragmatist and shared in a 2015 interview that he and others involved in submitting the petition had two goals: to raise public and government awareness of the decline of the monarch migration, and to generate funding for mitigation and conservation programs.[1] Though the ESA listing was denied in 2020, Brower's decision to support it fueled a monarch conservation flame that burns to this day.

▼

When the International Union for the Conservation of Nature (IUCN) declared in July 2022 that the migratory monarch butterfly had been added to its Red List of Threatened Species as "endangered," I couldn't help but think of Lincoln Brower. Surely the revered lepidopterist was cheering from the afterlife for conservationists to keep up the good fight.

The IUCN, founded in 1948 and based in Switzerland, is composed of government agencies and NGOs from around the world. It bills itself as "the barometer of life." The organization encourages a broad understanding of the pressures on worldwide life forms to guide best conservation efforts, but it has zero government authority or enforcement capacity. The IUCN attributed the monarchs' decline to climate change and habitat destruction.

Just as in 2014, after the ESA petition was filed, sensational headlines followed. The lead at National Geographic: "Monarch Butterflies Are Now an Endangered Species."[2] Members of the Monarchy—and the general public—were confused. Many had never heard of the IUCN. Naturally, some in the United States associated an "endangered" listing with their own government's Fish and Wildlife Service and scary rules about interacting with their favorite butterfly. Just two months earlier, the eastern migratory monarch population had shown an increase of 35 percent over the previous year. Five months before that, the western monarch population had vaulted more than 12,000 percent. Yes, 12,000 percent! Granted, that leap started from a historically low point, but still . . . wasn't that a hopeful sign?

In this context, the IUCN news came as a shock. Scientists and nonscientists weighed in. Migration studies expert Andy Davis bemoaned a "dogma of doom" that the listing promoted and within months even filed an official challenge to the 2022 decision.[3] For Davis, the listing made the public narrative of the monarch more and more divorced from reality, and he went so far as to label the listing a publicity stunt. Citizen scientists wondered aloud, "Can we still tag monarch butterflies?"

Chip Taylor questioned the scientific basis for the classification, in particular taking issue with the designation "migratory monarch butterfly" as a separate subspecies. In August 2023, Taylor challenged the IUCN listing in a Monarch Watch blog post titled "Why There Will Always Be Monarchs."[4] And Karen Oberhauser, who served on the IUCN committee that oversaw the listing, noted that, though an "endangered" red listing carries no regulatory authority, it serves as a "tool" meant to catalyze conservation action.[5]

A species needs to meet only one of five criteria to be eligible for IUCN red list "endangered" status: (A) the species' population decline, (B) the limitations of its geographic range, (C) whether the species exhibits a small population size, (D) whether the species lives in a restricted area, and (E) whether research indicates a high probability of extinction in the wild.[6] To assess a species, IUCN work groups of paid staff and volunteers gather in person or remotely to evaluate the candidate on the basis of the best available evidence. If a single one of the above criteria is met, the species is eligible for "endangered" status and is considered to be facing a "very high risk of extinction in the wild."

In the case of the monarch, criteria B through E were not applicable. The IUCN listed it under criterion A: "Population reduction observed, estimated, inferred, or suspected in the past where the causes of reduction may not have ceased OR may not be understood OR may not be reversible."[7] In the listing, the IUCN specified that *Danaus plexippus plexippus*, a subspecies known as the "migratory monarch butterfly," was endangered, while regular *Danaus plexippus*, the chosen designation of the science and nonscience community for monarchs, was not.

Taxonomy is highly contentious, and the scientific classification of species constantly changes. Even so, many monarch scientists balked at this hair splitting tactic to justify the listing. According to Chip Taylor, using the subspecies designation as an excuse to declare monarchs endangered made no sense. Even if we were to lose the migration, the subspecies would persist in Mexico and elsewhere, and the ability to migrate would not be lost for a "very, very long time." And since the subspecies itself is found widely in many other locations—Australia, New Zealand, Hawaii—it is not threatened with extinction.[8]

When asked about the subspecies designation, Monika Bohm, a cochair of the committee, referenced the *Butterflies of North America* website as the IUCN species resource but agreed that taxonomy is famously contentious. As for the jarring optics of listing a species as endangered when it had just experienced drastic population increases, Bohm said the large counts stumped the committee as well. She then cited the precautionary principle "It's better to be safe than sorry." Though the committee recognized that monarch numbers in 2022 "were really great, we didn't know if that was going to be a one-off. . . . It's better and less damaging to overstate and amend later, than to underestimate threats."[9]

But is it? In addition to boosting suspicion of science in an age of science denial, inflammatory rhetoric can backfire. It seems unfathomable to categorize the monarch population as "small" given that millions of the butterflies overwinter in Mexico each fall, and millions more occupy local populations or perish on the journey, with a significant number now winterbreeding in California and other warm states. Each hectare counted at Mexico's overwintering sites represents an estimated 21 million monarchs, begging the question, How can we justify "endangered" status for a species so abundant?

Talk of extinction precipitated the enactment of the previously mentioned "hands-off" policy in California, which resulted in a statewide mandate against touching, netting, rearing, or handling monarch butterflies in any way—even if found dead, even in one's own back yard—without a scientific collection permit. That translated into no tagging, no rearing, no testing for OE. No monarchs in the classroom, no harvesting them from gardens to witness the magic of their metamorphosis, unless you were associated with a university or research institution and had secured the appropriate license. The California Department of Fish and Wildlife adopted this "more cautious approach" in response to the dramatic drop in western monarch numbers in 2021.

Many involved in the Monarchy, including scientists, have serious concerns that forbidding people from interacting with monarchs rather than encouraging them to engage with and not fear them will lead to a world in which people neither understand nor care about them. This approach seems particularly inappropriate given the legacy of volunteers and citizen scientists who have made its natural history understood.

May 12, 2022: Armed California agents enforce the "hands-off" rule at the home of a volunteer researcher.

Researcher David James tells a story of getting into "big trouble" with state officials after recruiting three volunteer citizen scientists to help him count California's winter monarch population during the 2020–21 season. He and his collaborators ceased their research, which resulted in the first ever article to study winter breeding in the western monarch population.[10]

When James and his citizen scientist collaborators began their surveys in 2020, it was legal to interact with monarchs. Volunteers observed, counted, collected, and sometimes tagged and tested the creatures, all in an effort to tally the overwintering population and document changes in their behavior. But after the CDFW's ruling in mid-2021, such activity was declared illegal.[11] According to James, the CDFW seemed most upset by the tagging. He and his crew stopped tagging, but CDFW officials still sent weapon-toting officers to the front doors of his volunteers, demanding that they "not touch monarch butterflies." Footage of the armed officers was captured on a doorbell cam. "This is bad for bringing nature to people," said James. "We're creating children with no love of nature who just turn to their screens."

Since the enactment of the hands-off rule, citizen scientist Gail Morris, who serves as coordinator at the nonprofit Southwest Monarch Study from her home in Arizona, has had a tough time recruiting volunteers in California to assist with data collection. She has worked with hundreds of volunteers for more than a decade, work that resulted in a study published in 2015, "Status of *Denaus plexippus* Population in Arizona."[12] The crowdsourced research shed new insights on Arizona monarchs, finding that they migrate to both Mexico and California, are present in the state year round and breed there in the winter, and have very low levels of OE. Without the assistance of volunteers in California, the study would not have been possible.

Morris hesitated to complain, since she had just submitted the first request by a citizen science organization for a scientific permit and preferred not to call attention to herself. She expressed concern that many people are frightened of breaking the law and thus avoid doing anything. Over time, "a danger of indifference and helplessness sets in."[13]

▼

"Everyone is resigned to the fact that the migration will become a thing of the past," IUCN's Monika Bohm told me. It's true. Every monarch butterfly expert or scientist with whom I have spoken over the years believes the monarch butterfly migration is going away. It's not a question of if but when—and what we can do to manage and slow down the inevitable to preserve this spectacular biological phenomenon for as long as possible.

In September 2023, the IUCN revised its listing of monarch butterflies from "endangered" to "vulnerable." Bohm, now a member of the IUCN SSC Butterfly and Moth Specialist Group, explained that the monarch is "not out of the woods and is still a threatened species" and that IUCN intends to reassess the status of the species in five years' time."

David James's recent research paints a more adaptable, optimistic picture of the species than traditional narratives and news headlines suggest. Migration studies experts tell us that migrations are ephemeral. They serve a moment in time. When that time passes, creatures adapt. Creatures migrate because they have to—to find food, shelter, a hospitable climate, and partners for reproduction. If they can satisfy all those needs locally and achieve biological success by reproducing, why take on the enormous and often fatal challenge of migration?

As witnessed in California, monarch butterflies are increasingly adapting to warmer temperatures and available host plants, reproducing throughout the winter and boosting the number of total butterflies present along the West Coast. If western monarch butterflies can find a welcome place to overwinter, become reproductive, locate milkweed, and survive OE, drought, and atmospheric rivers, why should they migrate? Maybe eastern monarchs could do the same, becoming more like other seasonal butterflies, showing up at predictable times of the year, finding hospitable hideouts when the climate doesn't cooperate.

Meanwhile, a 2018 study led by Hannah Vander Zanden, an assistant professor at the University of Florida, showed that half of the monarch butterflies overwintering in Florida actually were migratory monarchs that originated in the midwestern breeding grounds. Using stable isotope analysis, a technique pioneered by Lincoln Brower, Vander Zanden and her team set about to determine the origins of monarchs found in the south Florida region. The team ground up samples of Floridian monarchs' wing tissues, ran them through appropriate equipment, and determined the insects' "isotopic signature"—that is, chemical clues that suggest where the insects grew up and the provenance of their milkweed.

The data showed that 40 percent of what were assumed to be local monarch butterflies actually came from elsewhere. They were migrants. Of those 40 percent, the majority started in the midwestern breeding grounds—not in Mexico—and some originated in Texas. In short, the study challenged general theories about monarch butterfly migration, particularly the singular importance of the massive migration to overwinter in the Mexican mountains, and found that monarch butterflies employ both migratory and resident life history strategies.[14]

As Andy Davis wrote when promoting the Vander Zanden article, "We have always assumed that the winter destination of the eastern breeding population is the mountains of Central Mexico. But what if it isn't? What if they don't ALL travel to Mexico? And . . . what if over time, greater and greater numbers of monarchs are choosing to travel to these 'alternate' winter destinations, like South Florida?"[15] And what if those moving up and down the West Coast begin winter breeding in urban gardens?

What if, indeed?

▼ ▼ ▼

CHAPTER 19

WELCOME HOME

It was a sunny November day, just a week after Día de los Muertos. A lone monarch butterfly sailed down the Avenida de Torres as we departed the Observatorio bus station on the west side of Mexico City. Mexico's monarch butterfly roosting sanctuaries would open to tourists the following week, but Bob and I had arranged to see them early.

Securing the special permit from CEPANAF, the state commission on natural parks and fauna, was no simple task. It required multiple emails, phone calls, text messages, guidance from Mexican friends, and the uncomfortable sharing of my birth certificate, passport, and utility bill via email with government officials in Zitácuaro. They seemed to regard me with suspicion. My *amigos mexicanos* suggested that a story I had previously written about the international conglomerate Grupo Mexico made them wary of my motives. The company was rumored to be moving quietly to reopen a mining operation in Angangueo, the heart of the monarch roosting sites' protected zone and not far from El Rosario, the most visited monarch sanctuary in Mexico.

After weeks of bureaucratic finagling, our permit was approved. This would be my first time to see the monarchs come home. Each of my three previous visits to the sanctuaries had occurred in the spring, when the insects begin their iconic months-long, multigenerational migration north.

The bus from Mexico City to Zitácuaro took two hours and ambled through the Mesa Central. Shallow pools still covered parts of the prehistoric lakebed and the Aztecs' ancient island capital city of Tenochtitlan. As we moved farther west, rows of rigid brown corn stalks lined little *milpas*, family-farmed plots along the roadside, their maize cobs harvested weeks earlier. Purple, white, and yellow flowers animated the countryside as our bus climbed the winding roads. Bob and I inhaled the scene from our seats just in front of the onboard restroom. A G-rated action flick blared from three video screens positioned on the overhead racks. Our new friend Ernesto, a six-year-old passenger sitting one row up, asked for help: could Bob help unscrew the stubborn cap on his liter of Coca-cola? *Por favor?*

Zitácuaro served as our base the first two nights. As we entered the gritty town, I looked for a motor court like the one Catalina described in her outings with Ken in their Winnebago almost a half-century earlier. I didn't see one. I also wondered where Bill Calvert and his troupe stayed during their travels after descending the mountain in search of roosting sites.

A friend from Morelia picked us up the next morning for the one-hour drive to El Rosario, which sits uphill from the small former mining town of Angangueo. Since my last visit years earlier, El Rosario had boomed into an even more crowded ecotourism destination for those seeking close encounters with monarch butterflies. Commerce huts crowded the parking lot, offering locally produced monarch butterfly souvenirs: embroidered napkins, *rebozos* (handwoven shawls), earrings, bracelets, all graced with our favorite orange-and-black insect. The car and bus plaza also featured a chapel and authentic local dining—delicious blue corn tortillas and chorizo verde, a green, cilantro- and chile-spiked sausage that rivaled the evergreen forest in its vibrancy.

Our guide, Manual Cruz Posadas, was lanky, agreeable, and fifty-nine years old. He led us on the hour-long hike up to 10,000 feet, answering my constant questions with good humor. Did the population size seem normal this year? Had he ever found a tag? Was he here for the great storm of March 2016? *Sí, sí,* and *sí,* said Cruz, all affirmative. What's the most interesting thing he had witnessed in the course of working as a guide at the sanctuary? He gazed into the distance and assumed a thoughtful look. A long moment passed before he responded. The chronically ill often request a visit to El Rosario as a last wish, he said. They arrive with cots, "*y los llevamos a ver a las mariposas*" (and we carry them up to see the butterflies). He stated this with obvious satisfaction and a smile, the silver cap on his front tooth glinting in the sun.

We continued up the relatively civilized trail, carved from the mountain and maintained by the community. The well-managed path makes the Llano de los Conejos ("Flat of the Rabbits") more accessible to tourists and is likely the reason El Rosario is the most popular of the preserves. Only slightly winded, we arrived at a grassy plain where the forest broke, offering an unobstructed view of monarch butterflies gliding against a bright blue sky. Pointing to the forest, Cruz said the monarchs were convening deeper in the woods. They were still in the early days of forming their winter roosts, he explained. When a cloud moved in front of the sun, the butterflies seemed to disappear into the forest. A folk-art mural at El Rosario's entrance accurately described their behavior: "When the sun shines, the monarchs fly; when it's cloudy, the monarchs rest."

Our trek to the tentative roosts took about fifteen more minutes. There they were—clustered like grapes, clinging to the trees, evoking sighs and whispers of relief and disbelief. We tried to capture their magic with our cameras and cell phones. Not possible. After a spell, we hiked back down the mountain feeling blissfully sated.

The next morning we vacated our room at the Casa de los Recuerdos in Zitácuaro and took a taxi to the village of Macheros. There our friend Ellen Sharp, a coowner of JM's Butterfly B&B, arranged for Vicente Moreno Rojas to guide us on an ambitious climb up Cerro Pelón. The bald hill, where Catalina and Ken first encountered the roosts back in 1975, is known these days as one of the more pristine of the sanctuaries open to the public.

We started on horseback, and it wasn't easy. A steep grade and the rocky, slippery trail—no guardrails or steps like El Rosario—made for an extremely challenging trek in this thin mountain air. My horse, Canela, so named for her cinnamon-colored coat, was the senior mare, very experienced, Vicente assured me. She and I led the way. Yet even she faltered at times, halting before large rocks and hesitating at splits in the path—right or left? Vicente prodded her with "Psht!" sounds. Since we were visiting preseason, the horses were a bit skittish and out of shape.

After an hour we arrived at another grand plain, this one the Llano de Tres Gobernadores ("Flat of the Three Governors"). The wide, grassy expanse divided two stands of oyamel and pine forest. A recent lightning strike had decimated several treetops and left one large conifer completely destroyed, its massive trunk felled like roadkill. Monarch butterflies perched on low-growing forbs, restoring themselves with sips of morning dew.

Vicente advised that in February, when the full population of eastern monarch butterflies resides here, this plain is a *tapeta de mariposas*, a carpet of butterflies.

We took a break for a picnic packed by Moreno's sister—ham sandwich, chips, apple and, *pedacito de chocolate*, a small bite of chocolate. Francisco Moreno Hernandez, an arborist for Butterflies and Their People, AC, a Mexican nonprofit started by Sharp and Joel Moreno to protect the forest and its inhabitants, approached us on horseback. Francisco advised that the butterflies were forming roosts farther up the mountain, above 11,000 feet. It's about forty-five minutes, a difficult hike, he said. A forest ranger patrolling the area ignored a local family joy riding on a four-wheeler and joined our conversation. He was curious about the presence of preseason gringos. After introductions and a brief discussion that included the inspection of my permit, he agreed to lead us to the roosts.

We started again on horseback but dismounted when the narrow, rocky trail made the ride impossible. Canela seemed relieved. Relatively fit for my sixty-one years, I had never endured such a literally breathtaking hike. I panted like a dog, breathing hard with each step. Even the horses were lathered with a sheen of sweat on their muscular flanks. Every few minutes, I stopped to absorb the captivating sight of an increasing number of monarchs flitting above. I paused to find a foothold, gazed over the treetops, and thought of Catalina. How did she ever find her way to this absurdly remote and majestic place? Ah yes, she was twenty-four years old and had recruited a local the day of her momentous discovery.

Vicente kept asking me if I was okay, and I assured him that yes, I was fine. But my heart banged hard in my chest and I leaned full tilt into the walking stick the forest ranger slashed for me with his machete. "*Casi estamos? Are we almost there?*" I asked him, the proverbial back-seat child on the interminable road trip.

"*Quince minutos mas.*" Fifteen more minutes.

Bob was up ahead. Just months ago he had endured major back surgery—they literally sliced open his spine and inserted cushions between his fourth and fifth vertebrae, what he called his "titanium Legos." Now he was a week from his sixty-fifth birthday. I was impressed with his stamina. "You okay?" I called out to him. He was fine, too.

The author witnesses the monarchs return "home."
Photo by Robert Rivard.

We arrived at a crest and took a moment to appreciate the stunning mountains surrounding us. The scent of pine trees hung in the air. Monarchs glided through the forest and across the sky like daytime's shooting stars. We had to be getting close.

This slow, arduous trudge reminded me of Fred Urquhart's historic 1976 *National Geographic* account. The Canadian scientist had chased and studied the monarch migration for more than four decades. At the age of sixty-five, not far from here and only months older than my husband, he had struggled physically to finally witness the subject of his obsession. In the cover story, he described how he and his wife Nora walked along the mountain crest, their hearts pounding, feet leaden. "The rather macabre thought occurred to me: Suppose the strain proved too much?" he wrote.

Catalina and Ken had arranged for the couple to use oxygen tanks on their hike. Now I understood why.

And what about the unpleasant incident in the forest? The two most accomplished monarch butterfly scientists of the twentieth century, locked in their stubborn, testosterone-fueled rivalry. They ran into each other on these inaccessible mountaintops. What are the chances?

Another ten minutes and we arrived. Half a dozen sacred firs reached for the skies. Chilly butterflies shivered on the ground, vibrating their wings to warm up for flight. Occasional torn wings littered the forest mulch, suggesting a meal consumed by birds or lizards. Tens of thousands, perhaps hundreds of thousands, of monarch butterflies crowded the oyamel branches. Some huddled on the trees' massive trunks. Others wandered with seeming purpose across the whitewashed sky, seeking a safe place to be with their kind.

I found my own spot just yards from the roosts. I sat on the rocky earth, removed my hat, leaned back on the stump of a felled oyamel tree, and savored the majestic natural spectacle. I gazed at the lilting creatures and marveled at the reassuring reality that somehow each one had found its way. Hopefully, we can, too.

▼ ▼ ▼

AFTERWORD

I've made several springtime visits to the roosting sites in Mexico since my singular autumn trip to see the monarchs return "home" in the fall of 2017. Each time, I'm flooded with conflicted feelings and swear it will be my last. Yes, ecotourism helps pay the bills. And yes, watching millions of monarch butterflies lilt through the thin mountain air reinforces the magic and boosts appreciation of nature's wonders.

But it's impossible to not be alarmed by the overwhelming negative influence of humanity on these precious mountain roosts. Every visit provokes justified guilt for the carbon footprint, the erosion of the mountain, the commercializing of nature, and the disrespectful behavior human beings display in this sacred setting.

In 2020, when I last visited El Rosario, the price to witness this unique natural spectacle was less than the cost of a mediocre latte at your local coffee bar. Admission to Mexico's most popular monarch butterfly sanctuary: 50 Mexican pesos—US $2.77. Seeing these magical creatures cluster like grapes in the towering firs or flush in a butterfly wave on a sunny winter day is a bucket list dream for many people. The cost of admission should better reflect how special this is, because, as we know, it won't last forever.

At El Rosario, hundreds of people crowd the dirt path, jockeying for position to snap photos and capture the moment. The sanctuary has worked to appease growing crowds. Four sprawling asphalt parking lots skirt the once understated entrance to accommodate cars and tour buses. A separate paved trail for sightseers on horseback has been installed to facilitate quick transport to the butterfly viewing area. Walking the steep path, which starts at an elevation of 10,000 feet, takes about thirty minutes but requires navigating more than 250 food and souvenir booths that precede the entrance.

A ticket to Mexico's most popular monarch butterfly
sanctuary should cost more than $2.77.
Photo by Monika Maeckle.

Inside, at Llano Cruzado, where five dirt roads intersect, foot-deep tire ruts scar the earth, a consequence of clearing diseased trees. According to our guide, each August the ejidatarios collectively determine which trees should be removed. Bark beetles are a growing problem as the climate warms, so damaged trees are removed to tidy up the forest before the butterflies arrive. Yet the gashes from the semi used to haul away the lumber persist, as will the erosion and fragmentation of the forest.

At one point, visitors crowd a muddy trail crossing, as a seep of water dampens the soil and the insects puddle in enchanting waves of orange and black. Three young women squat to take turns holding a dead butterfly retrieved from the ground. "Ok, you go," one says to the others in Spanish, propping the butterfly corpse on her pointer finger in a quest for selfies.

Foreigners could easily and willingly pay more to enter the sanctuaries; limits could be set on the number and pace of people allowed access, which would boost the ecotourism experience and the local economy. Conservationists have lobbied for years to limit access, to no avail. Cuzco, Peru, implemented such restrictions in 2017. In 2023 access to the ancient Inca fortress Machu Picchu cost about \$45, with visitors limited to 2,500 per day. Reservations are required.

I often wonder what a monarch butterfly would say if we could interview her. What would she think of our insistence that she continue an arduous, often fatal migration rather than become a local if conditions are hospitable? How would she view our "need" to continue this grand natural spectacle so our offspring could witness it? How would she feel about being snagged in midflight en route to Mexico and having a tag or radio transmitter attached to her torso so we could record her status?

As of this writing, the U.S. Fish and Wildlife Service is scheduled to reconsider listing the monarch butterfly as threatened under the Endangered Species Act sometime in 2024. If it lists the monarch, our relationship with this charismatic insect will be redefined. Engagement and interaction with it will be practically verboten. If the monarch remains unlisted, conservation efforts will continue under the shadow of a potential listing that will be reconsidered every two years.

I embrace the optimism of the California studies and other research that indicates monarchs and other butterflies are highly adaptable and are morphing to their next evolutionary phase. I'm hoping they will become more like other Lepidoptera—showing up seasonally, more sporadically, and in numbers that vacillate between few and vast. They will still be present and captivating. Perhaps they can emulate the life cycle of another orange-and-black butterfly that moves up and down the Interstate Highway 35 pollinator corridor, clogging windshields and car grills along the way when favorable conditions prevail. The American snout, *Libytheana carinenta*, a long-nosed butterfly with mottled black, orange, and white coloration, has a reputation for "mass movements" in and around Central and South Texas after late-summer rains. Like the monarch, snouts sip nectar from a variety of plants but lay eggs only on a specific host—in this case the hackberry, which is often viewed as an undesirable shrub but is actually a fantastic wildlife plant.

Every few years we get a "snout invasion" in San Antonio. Typically this happens when heavy rains follow a long dry spell, encouraging rapid, tender growth of hackberries that have gone temporarily dormant in response to drought. Like their host plant, snouts retreat into a dormant stage during dry periods, awaiting rain and more favorable conditions to be biologically successful. When the rains hit, they become activated, mate, and lay eggs. The result is often a massive presence of snouts.

As University of Texas integrative biologist Larry Gilbert put it, "When you see a big bullseye over South Texas brush country with four-five inches of rain? Ok, you'll be breathing snout butterflies in a month."[1] Gilbert believes we'll see more frequent snout invasions than in the past. "It used to happen every eight years or so," he said. "Now we're seeing this pattern every two-three years."

In September 1921, after a record downpour in Central Texas of 36.4 inches of rain in an eighteen-hour period, a snout butterfly breakout resulted. An estimated 25 million per minute southeasterly-bound snout butterflies passed over a 250-mile front from San Marcos to the Rio Grande. Scientists noted at the time that the butterflies' flight lasted an estimated eighteen days and may have involved more that 6 billion butterflies.[2]

▼

Human nature leads us to cling to what we know and to assume that's what should always be. Those who witnessed the "snout-break" of 1921 probably never forgot the incident's magnitude, measuring every year thereafter against the one they saw firsthand. But life cycles are up and down, and change is the only constant. Yes, losing the migration will be sad, but, as monarchs teach us, loss is part of the life cycle.

My German mother, who as a child during World War II endured long bouts in the bomb shelter, often repeated what her mother said to her when faced with difficult times. "This too, shall pass." So will the monarch migration. Monarchs will still exist. They'll show up seasonally and sometimes in huge numbers like their butterfly cousins the snouts. They'll still charm and enchant us. The great migration of millions of insects finding their way "home" to a place they've never been will evolve into a story of adaptation, a lesson we should all embrace.

▼ ▼ ▼

ACKNOWLEDGMENTS

I extend my most heartfelt thanks to University of Oklahoma Press Editorial Director Andrew Berzanskis. Were it not for his chancing upon a few chapters of my book on the Texas Writers' League website (as a finalist in the nonfiction writing category in 2022), I would likely still be shopping for an agent and/or editor and you would not be reading this. This project began in 2017 and I was beginning to lose faith when Andrew stumbled upon a few chapters, recognizing the light in its research and writing. For that I am extremely grateful. Gracias, Andrew, for believing in me, and this book.

Thanks, also, to the hundreds of scientists, citizen scientists, butterfly breeders, friends, and Texas Butterfly Ranch readers who have contributed to this curiosity tour. Your support and interest in my questions and answers have fueled and sustained me. Special shoutouts to Chip Taylor, who has been especially patient, tolerating my sometimes irritating questions via phone calls, texts, and emails, and my monarch mentor Jenny Singleton, who introduced me to the captivating creatures.

My husband, Bob Rivard, and sons Nicolas and Alex, as well as Cacteye, Brisket, Trouble, and Cocoa, have made this journey fun and possible. You are my ecosystem. Gracias to all.

NOTES

Introduction

1. Francisco Sánchez-Bayo,"Worldwide Decline of the Entomofauna: A Review of Its Drivers," *Biological Conservation* 232 (April 2019), 8–27.

Chapter 2

1. "The Monarch Butterfly in North America," U.S. Department of Agriculture/U.S. Forest Service, n.d., www.fs.usda.gov/wildflowers /pollinators/Monarch_Butterfly/index.shtml.
2. "Life Cycle," Monarch Joint Venture, n.d., https://monarchjointventure .org/monarch-biology/life-cycle.
3. Rothschilde cited in Lincoln P. Brower et al., "Monarch Sex: Ancient Rites, or Recent Wrongs?" *Antennae* 31, no. 1 (2007), 12–18. An account of the sex life of the monarch butterfly was given at the RES National Meeting, Bath University, September 22, 2006, sponsored by the RES, British Ecological Society, Linnean Society, and Natural History Museum: www .fzi.uni-freiburg.de/pdf/2007Brower_etal.pdf.
4. Anurag Agrawal, *Monarchs and Milkweed: A Migrating Butterfly, a Poisonous Plant, and Their Remarkable Story of Coevolution*, 1st ed. (Princeton, NJ: Princeton University Press, 2017), 63–89.

Chapter 3

1. Fred Urquhart, "Found at Last: The Monarch's Winter Home," *National Geographic*, August 1976, 161.
2. Interview with Catalina Trail, Texas Butterfly Ranch, July 10, 2012, https://texasbutterflyranch.com/2012/07/10/founder-of-the-monarch -butterfly-roosting-sites-in-mexico-lives-a-quiet-life-in-austin-texas.
3. Interviews with Catalina Trail, 2012–2016 by phone, email, and in person. It seems odd that no one claims to have seen the millions of butterflies. Catalina speculated that locals typically were suspicious and didn't want to reveal that they had; who knew what the gringo and his girlfriend had in mind? Maybe they would steal or ruin the monarchs' secret hideaway, like so many other Mexican treasures lost to conquistadors.

Chapter 4

1. Sue Halpern, *Four Wings and a Prayer*, 1st ed. (London, Weidenfeld and Nicolson, 2001), 89.
2. Screen shot from minute 16:20 of "Lincoln Brower Monarch Butterflies Interview," University of Florida Biography, Life Overview (2013), www .youtube.com/watch?v=g8Vrzd_SvTc.
3. Cristopher Koehler, oral history with Lincoln Brower, University of Florida Digital Collections, March 14, 1994, 3, https://original-ufdc.uflib .ufl.edu/UF00006168/00001/pdf.
4. The Browers fed monarchs toxic milkweeds and convinced others, after much trial and error, to consume cabbage leaves. The bird gobbled up the cabbage-fed monarchs, and when it was slipped one that fed on bitter milkweed it retched violently. The blue jay avoided monarch butterflies thereafter. Thus, the toxic, bitter cardiac glycosides from the milkweeds monarchs consume as caterpillars make their long migration possible in the sense that predators avoid them.
5. Lincoln Brower, phone interview with Monika Maeckle, June 6, 2017.
6. Lincoln Brower, "Understanding and Misunderstanding the Monarch Migration," *Lepidopterists' Society Newsletter* 49, no. 4 (1995), 333.

7. Koehler, oral history with Lincoln Brower, 21.

8. Urquhart, "Found at Last," 161.

9. Brower, "Understanding and Misunderstanding," 334–35.

10. Lincoln Brower, phone interview with Monika Maeckle, June 6, 2017.

11. Koehler, oral history with Lincoln Brower, 22.

12. Koehler, oral history with Lincoln Brower.

13. William Calvert, phone interview, June 2017.

14. Koehler, oral history with Lincoln Brower, 24.

15. Monika Maeckle, "Historic Rendezvous of Those Who Located Monarch Butterfly Roosting Sites Draws Crowd of 200," Texas Butterfly Ranch, March 28, 2014, https://texasbutterflyranch.com/2014/03/28/almost-200 -gather-for-historic-rendezvous-of-monarch-butterfly-roosting-site -founders.

16. Koehler, oral history with Lincoln Brower, 23.

17. Bayard Webster, "2d Group Uncovers Butterflies' Secret: Report on Monarchs' Winter Home Spurs Fear of Ecological Threat and Rivalry between Experts," *New York Times*, May 19, 1977, 21.

18. Lincoln Brower, phone interview June 6, 2017.

19. Alec Scott, "Where Do You Go, My Lovelies? Fred and Norah Urquhart's Lifelong Quest for the Hidden Kingdom of the Monarch Butterfly," *Toronto Magazine*, August 24, 2015, http://magazine.utoronto.ca/blogs /where-do-you-go-my-lovelies-norah-and-fred-urquhart-monarch -butterfly-migration.

20. Fred Urquhart, *The Monarch Butterfly, International Traveler* (Chicago: Nelson Hall, 1987), 190.

21. Halpern, *Four Wings and a Prayer*, 93.

22. *People Magazine*, April 3, 1978.

Chapter 5

1. Email correspondence with *Natural History*, Mai Reitmeyer, Sr. Research Services Librarian, March 7, 2017.

2. Donald A. Davis, "Fred Urquhart (1912–2002)," Entomological Society of Canada, 2003, obituary of Fred Urquhart, http://esc-sec.ca/wp/wp -content/uploads/2017/02/Obit_Urquhart_Fred.pdf.

3. "Chasing Monarchs on Horseback," *Insect Migration Studies Newsletter,* March 1965, 4, https://monarchwatch.org/read/articles/index.htm.

4. Fred and Norah Urquhart, Insect Migrations Studies Annual Report to Research Associates, 1994, 3:2.

5. "Monarca, Espiritu de la Bosque," Milenio documentary, 2016, www.youtube.com/watch?v=GkUSxnCpBYI.

6. Milenio press release, November 29, 2018.

7. Interview by phone with Karen Oberhauser, March 13, 2017.

8. Monarch Joint Venture Annual Report, 2016.

9. Adam Federman, "All Aflutter: The Flap over the Mail Order Butterfly Industry," *Earth Island Journal,* September 2008, www.earthisland.org/journal/index.php/eij/articlc/all_aflutter.

10. Federman, "All Aflutter."

11. Karen Oberhauser, Monarch Watch D-Plex list post, March 8, 1996, www.monarchwatch.org/read/articles/trip.htm.

12. In October 2023, Oberhauser announced her intent to retire. She planned to stay involved in monarch science and expressed excitement about taking a chain-sawing class for women to make her a more well rounded land steward.

Chapter 6

1. The vocabulary continues to evolve. In an effort toward inclusion, some organizations in recent years have recast "citizen science" as "community science." Caren B. Cooper, "Inclusion in Citizen Science: The Conundrum of Rebranding: Does Replacing the Term 'Citizen Science' Do More Harm Than Good?" *Science,* June 25, 2021, www.science.org/doi/abs/10.1126/science.abi6487.

2. Chip Taylor, Monarch Watch seasonal summary, April 21, 1993, University of Kansas at Lawrence, 1.

3. Rex Dalton, "Lab Allergies Force Some Scientists to Take Cover or Change Careers," *Scientist,* November 11, 1990, www.thc-scientist.com/news/lab-allergies-force-some-scientists-to-take-cover-or-change-careers-60993.

4. Phone, email, and in-person interviews with Chip Taylor, 2011–2017.

5. Monika Maeckle, "The Mark of a Monarch Tag," *Two Million Blossoms*, Summer 2020, 80.

6. Chip Taylor, Monarch Watch seasonal summary, May, 1996 University of Kansas at Lawrence, 12.

7. Interviews with Andre Green, 2022–2023.

Chapter 7

1. Gail Morris, Karen Oberhauser, and Lincoln Brower, "Estimating the Number of Overwintering Monarchs in Mexico," Monarch Joint Venture, December 6, 2017, https://monarchjointventure.org/blog/estimating -the-number-of-overwintering-monarchs-in-mexico#:~:text=To%20 date%20the%20west%20population,a%20lack%20of%20complete%20 data.

2. Kate Galbraith "Governor Declares Days of Prayer for Rain," *Texas Tribune*, April 21, 2011, www.texastribune.org/2011/04/21/texas-governor -declares-weekend-of-prayer-for-rain.

3. John W. Nielsen-Gammon, "The 2011 Texas Drought: A Briefing Packet for the Texas Legislature," October 31, 2011, 16, https://core.ac .uk/download/pdf/79654057.pdf.

4. Lucy Hicks, "We Need a Better Way to Measure Monarch Populations: An Iconic Butterfly Is in Trouble—but How Much Trouble?" *Science Line*, April 4, 2018, https://scienceline.org/2018/04/need-better-way-measure -monarch-populations.

5. Wayne E. Thogmartin et al., "Density Estimates of Monarch Butterflies Overwintering in Central Mexico," National Library of Medicine, National Center for Biotechnology Information, April 26, 2017, www .ncbi.nlm.nih.gov/pmc/articles/PMC5408724.

6. CBS News, January 29, 2014, www.cbsnews.com/news/monarch -butterflies-drop-migration-may-disappear; Michael Wines, "Migration of Monarch Butterflies Shrinks Again under Inhospitable Conditions, *New York Times*, January 29, 2014, www.nytimes.com/2014/01 /30/us/monarch-butterflies-falter-under-extreme-weather.html#:~:text =The%20migrating%20population%20has%20become,in%20danger%20 of%20effectively%20vanishing.

7. Elise Cohen, "Spring Has Sprung: The Sixth-Annual White House Garden Planting," White House brief, April 3, 2014, https://obamawhitehouse.archives.gov/blog/2014/04/03/spring-has-sprung-sixth-annual-white-house-garden-planting.

8. Juliet Elperin, "Michelle Obama Makes the Pitch for Pollinators," *Washington Post*, June 3, 2015, www.washingtonpost.com/news/post-politics/wp/2015/06/03/michelle-obama-makes-the-pitch-for-pollinators/?utm_term=.6e68bob77c04.

9. Office of the White House Press Secretary, "Presidential Memorandum—Creating a Federal Strategy to Promote the Health of Honey Bees and Other Pollinators," June 20, 2014, https://obamawhitehouse.archives.gov/the-press-office/2014/06/20/presidential-memorandum-creating-federal-strategy-promote-health-honey-b.

10. National Pollinator Strategy to Promote Health of Honeybees and Others, May 19, 2015, Pollinator Task Force, https://lccnetwork.org/sites/default/files/Resources/pollinator_health_strategy_2015.pdf.

11. Oberhauser interviewed by Monika Maeckle, "Monarch Butterfly Population up in Mexico, down in California," Texas Butterfly Ranch, January 17, 2019, https://texasbutterflyranch.com/2019/01/17/monarch-butterfly-population-up-in-mexico-down-in-california.

12. Matthew Renda, "California Drought-Free for First Time in 8 Years," *Courthouse News Service*, March 14, 2019, www.courthousenews.com/california-drought-free-for-first-time-in-8-years.

13. Sarina Jepsen and Scott Hoffman Black, "Western Monarch Population Closer to Extinction: Still No Federal or State Protection in Sight," Xerces Society, January 19, 2021, https://xerces.org/press/western-monarch-population-closer-to-extinction-still-no-federal-or-state-protection-in-sight; Isis Howard and Emma Pelton, "Western Monarch Thanksgiving Count Tallies Nearly 250,000 Butterflies," Xerces Society, January 24, 2022, https://xerces.org/blog/western-monarch-thanksgiving-count-tallies-nearly-250000-butterflies#:~:text=The%202021%20Thanksgiving%20Count%20total,of%201ess%20than%2030%2C000%20monarchs.

Chapter 8

1. Agrawal, *Monarchs and Milkweed*, 22–42.
2. Columba González-Duarte, "Resisting Monsanto: Monarch Butterflies and Cyber-Actors," in *Resistance to the Neoliberal Agri-food Regime: A Critical Analysis*, ed. Steven Wolf and Alessandro Bonanno, 165–79 (Abingdon: Routledge, 2018).
3. González-Duarte, "Resisting Monsanto."
4. Andy Davis, "New Study Published: Despite Winter Colony Declines, Monarchs Are Thriving in North America (Really)," June 10, 2022, www.monarchscience.org/single-post/new-study-published-despite -winter-colony-declines-monarchs-are-thriving-in-north-america -really.
5. Emily Anthes, "Some Monarch Butterfly Populations Are Rising: Is It Enough to Save Them?" *New York Times*, June 10, 2022, www .nytimes.com/2022/06/10/science/monarch-butterflies-mexico.html ?searchResultPosition=4.
6. Chip Taylor, "Evaluating the Migration Mortality Hypothesis Using Monarch Tagging Data," *Frontiers in Ecology and Evolution*, August 7, 2022, www.frontiersin.org/articles/10.3389/fevo.2020.00264/full.
7. In response to the Taylor study, three biologists challenged some of his methodologies and labeled the "milkweed limitation hypothesis" versus "migration limitation hypothesis" debate a false dichotomy: James A. Fordyce, Chris C. Nice, and Matthew L. Forister, "Commentary: Evaluating the Migration Mortality Hypothesis Using Monarch Tagging Data," *Frontiers in Ecology and Evolution* 8 (November 30, 2020), www .frontiersin.org/articles/10.3389/fevo.2020.604914/full.

Chapter 9

1. "Spread Your Wings and Fly Seminar," The Butterfly Website, September 7, 2002, www.butterflywebsite.com/series.htm.
2. Wehling retired in June 2022.
3. Robert Michael Pyle, "Under Their Own Steam: The Biogeographical Case against Butterfly Releases," *News of the Lepidopterists' Society* 52,

no. 2 (Summer 2010), http://washingtoninvasivespecies.weebly.com
/uploads/9/5/9/1/9591784/pyle_butterfly_release_article._pdf.pdf.

4. Phone interview with Robert Michael Pyle, February 26, 2019.

5. "Petition Seeking Regulation of Bumble Bee Movement," Xerces Society blog, January 12, 2010, https://xerces.org/publications/petitions
-comments/petition-seeking-regulation-of-bumble-bee-movement
-january-12–2010.

Chapter 10

1. Various interviews with Catalina Trail via mail, phone, and in person, 2012–2023.

2. I asked Catalina in 2023 if I could have formal permission to use one of the photos for this book, but she declined.

3. Interview with Carlos Gottfried, March 3, 2017, who started Monarca, the first *associación civil* (Mexican nonprofit organization) devoted to conservation.

4. "Flight of the Butterflies Wraps One-Year Shoot, Mexican President Visits Set on Last Day of Filming," SK Films, May 6, 2012, http://skfilms
.ca/flight-of-the-butterflies-wraps-one-year-shoot-mexican-president
-visits-set-on-last-day-of-filming.

5. Chip Taylor, "In Pursuit of a Little History," Monarch Watch 1998 Season Summary, May 1999, 24–25, https://monarchwatch.org/download/pdf
/season-summary-1998.pdf.

6. Chip Taylor, "In Pursuit of a Little History: A Retraction," Monarch Watch Blog, December 5, 2016, https://monarchwatch.org/blog/2016/12
/05/in-pursuit-of-a-little-history-a-retraction.

Chapter 11

1. Carol Kaesuk Yoon, "Festive Release of Butterflies Puts Trouble in the Air," *New York Times*, September 15, 1998, www.nytimes.com/1998/09
/15/science/festive-release-of-butterflies-puts-trouble-in-the-air.html.

2. Phone interview with Jeffrey Glassberg, January 9, 2019.

3. Beginning in 2023, Shady Oak Butterfly Farm transitioned to offering plants for native butterfly habitats and no longer raised or sold butterflies.

4. Kylie Mar, "Epic-Fail Butterfly Stunt Takes a 'Nasty' Turn on 'RuPaul's Drag Race' Season 10 finale," Yahoo Entertainment, June 18, 2018, www .yahoo.com/entertainment/butterfly-stunt-takes-nasty-turn rupauls -drag-race-season-10-finale-014847657.html.

5. Emma Pelton, "Keep Monarchs Wild: Why Captive Rearing Isn't the Way to Help Monarchs," Xerces Society, June 15, 2023, www.xerces.org /blog/keep-monarchs-wild.

6. "Late Blooming Butterfly Gets Airline Lift to Join Migration," *San Antonio Express News*, November 5, 2011, www.expressnews.com/living _green_sa/article/Late-blooming-butterfly-gets-airline-lift-to-join -4009850.php#photo-3694238.

7. "New Statement from Monarch Conservation Groups Says—For the Love of God, Stop Mass-Rearing," Science of Monarch Butterflies blog, September 11, 2018, http://akdavis6.wixsite.com/monarchscience/single -post/2018/09/11New-statement-from-monarch-conservation-group-says ---For-the-love-of-God-stop-mass-rearing-monarchs-in-your-kitchen.

8. As of this writing, The Beautiful Monarch Page was "paused" on Facebook in September of 2023 because of contentious comments. The administrator of the page may or may not reactivate it at any time.

9. Big Tree Butterflies closed in 2017 when Hurricane Harvey passed right over the operation's greenhouse.

10. The migratory population for the 2015–16 season occupied 4.01 hectares in Mexico. If we multiply that by the newer 21.1 million per hectare, we get 84,600 butterflies.

Chapter 12

1. Susan Mahr, "Tropical Milkweed, *Asclepias currasavica*," University of Wisconsin, Division of Extension, n.d., https://hort.extension.wisc.edu /articles/tropical-milkweed-asclepias-currasavica/#:~:text=With%20 numerous%20common%20names%20including,America%2C%20 Central%20America%20or%20Mexico.

2. John E. Abbot, *The Natural History of the Rarer Lepidopterous Insects of Georgia: Including Their Systematic Characters, the Particulars of Their Several Metamorphoses, and the Plants on Which They Feed* (London: T. Bensley, for J. Edwards, Cadell and Davies, and J. White, 1797), 10–11, www.biodiversitylibrary.org/item/5968#page/50/mode/1up.

3. These observations taken from Dara Satterfield, John C. Maerz, and Sonia Altizer, "Loss of Migratory Behaviour Increases Infection Risk for a Butterfly Host," Biodiversity Heritage Library, paper supplementary materials, February 22, 2015, http://rspb.royalsocietypublishing.org /content/282/1801/20141734.

4. Email exchange with Jim Nau, Ball Seed Company; John Dole, "1994 ASCFG National Cut Flower Trials," North Carolina State University, n.d., https://cutflowers.ces.ncsu.edu/cut-flower-cultivar-trials/1994 -cultivar-trials.

5. Thierry LeFevre, "Evidence for Trans-generational Medication in Nature," *Ecology Letters* 13, no. 12 (December 2010), 1485–93, https:// pubmed.ncbi.nlm.nih.gov/21040353.

6. LeFevre, "Evidence for Trans-generational Medication."

7. Interview with Lincoln Brower, Texas Butterfly Ranch, February 16, 2015, http://texasbutterflyranch.com/2015/02/16/q-a-dr-lincoln-brower -talks-ethics-endangered-species-milkweed-and-monarchs.

8. Jeffrey Glassberg, "Tropical Milkweed," *American Butterflies*, Winter, 2014.

9. "USDA revises plant hardiness zone map, seed packets will never be the same," Texas Butterfly Ranch, November 20, 2023, https://texasbut terflyranch.com/2023/11/20/usda-revises-plant-hardiness-zone-map- seed-packets-will-never-be-the-same/.

10. Glassberg, "Tropical Milkweed."

11. Personal interviews and email exchanges with Edith Smith, March 2018.

12. Christen Steele, Isabella G. Ragonese, and Ania A. Majewska, "Extent and Impacts of Winter Breeding in the North American Monarch Butterfly," *Current Opinion in Insect Science* 59 (October 2023), 101077, https://doi .org/10.1016/j.cois.2023.101077.

13. Ania A. Majewska et al., "Parasite Dynamics in North American Monarchs Predicted by Host Density and Seasonal Migratory Culling," *Journal of Animal Ecology* 91 (2022). 780–93, https://doi.org/10.1111/1365-2656.13678.

Chapter 13

1. "GM Foods Not Served in Monsanto Cafeteria," CBC News, December 22, 1999, www.cbc.ca/news/world/gm-foods-not-served-in-monsanto-cafeteria-1.173403.

2. The web page that hosted this document, https://monsanto.com/company/media/statements/monsanto-cafeteria, was retired after Bayer acquired the company in 2018.

3. "What We Do," National Fish and Wildlife Foundation, n.d., www.nfwf.org/what-we-do#:~:text=Since%20our%20creation%20by%20Congress,for%20current%20and%20future%20generations.

4. Milenio series: www.youtube.com/watch?v=8WZvYAXNZGI&t=56s. Other sponsors listed: OHL Mexico, Atleata S. A., and Tourism Council of Michoacan, the Mexican government.

5. Carey Gillam, *Whitewash: The Story of a Weed Killer, Cancer, and the Corruption of Science*, 1st ed. (Washington, D.C.: Island Press, 2017), 45.

6. C. M. Benbrook, "Impacts of Genetically Engineered Crops on Pesticide Use in the U.S.—the First Sixteen Years," *Environmental Sciences Europe* 24, no. 24 (2012), https://enveurope.springeropen.com/articles/10.1186/2190-4715-24-24.

7. Bill Freese and Marth Crouch, "Monarchs in Peril: Herbicide-Resistant Crops and the Decline of Monarch Butterflies in North America," Center for Food Safety executive summary, February 2015, www.centerforfoodsafety.org/files/monarch_es_82936.pdf.

8. Robert G. Hartzler "Reduction in Common Milkweed (*Asclepias syriaca*) Occurrence in Iowa Cropland from 1999 to 2009," *Crop Protection Journal* 29, no. 12 (December 2010), 1542–44, www.sciencedirect.com/science/article/pii/S0261219410002152.

9. Greens European Free Party Alliance press release, December 5, 2016, https://www.greens-efa.eu/en/article/news/we-are-pissed-off.

10. Reuters, "Monsanto Ordered to Pay $289 Million in Roundup Cancer Trial," *New York Times*, August 10, 2018, www.nytimes.com/2018/08/10/business/monsanto-roundup-cancer-trial.html#:~:text=The%20company%20was%20ordered%20pay,glyphosate%2Dbased%20weedkillers%20cause%20cancer.

11. Anne Barnard, "N.Y.C. Bans Pesticides in Parks with Push from Unlikely Force: Children," *New York Times*, April 24, 2021, www.nytimes.com /2021/04/24/nyregion/pesticide-ban-nyc.html?searchResultPosition=1.

12. Bayer Neonicotinoids Report, 2022, www.bayer.com/sites/default/files /bayer-neonicotinoids-report-2022-sngl.pdf.

13. Phone interview with Pamela Bachman, January 27, 2023.

14. Monsanto news release, https://monsanto.com/news-releases/continued -soybean-technology-expansion-and-cost-discipline-expected-to-drive -return-to-eps-growth-in-fy17.

15. Bayer's Monarch Flyer, www.bayer.com/sites/default/files/2022–03/Bayer -AG-Financial-Statements-2021.pdf.

Chapter 14

1. Matthew Bremner, "Did Avocado Cartels Kill the Butterfly King?" *Bloomberg News*, December 7, 2021, www.bloomberg.com/news/videos /2021–12–07/did-avocado-cartels-kill-the-butterfly-king-video.

2. Ernesto Martínez Elorriaga, "Michoacán: No Trace of Homero Gómez, Missing Forestry Activist," *La Jornada*, January 27, 2020, https:// lajornadasanluis.com.mx/nacional/michoacan-sin-rastro-de-homero -gomez-activista-forestal-desaparecido.

3. The CEC was formed in 1994 to implement the North American Agreement on Environmental Cooperation (NAAEC), the environmental side of the North American Free Trade Agreement.

4. Brook Larmer, "How the Avocado Became the Fruit of Global Trade," *New York Times Magazine*, March 27, 2018, www.nytimes.com/2018 /03/27/magazine/the-fruit-of-global-trade-in-one-fruit-the-avocado .html#:~:text=The%20United%20States%20had%20banned,An%20 avocado%20explosion%20followed.

5. Agricultural Marketing Resource Center, www.agmrc.org/commodities -products/fruits/avocados.

6. Avocado Institute of Mexico Fact Sheet 2022, January 2023, https:// avocadoinstitute.org/avo-economics/avo-economics-fact-sheet.

7. Avery Haines, "Narco Avocados: Violent Drug Cartels Are Taking Over Mexico's Avocado Industry," W5 investigation, Canadian

Television CTV, February 2, 2023, www.youtube.com/watch?v
=116VGMjzs10.

8. Columba González-Duarte and Manuel Ureste, "Indigenous Communities in Mexico Take Up Arms to Defend the Monarch Forest," NACLA,
March 21, 2021, https://nacla.org/mexico-monarchs-organized-crime.

9. Submission Update, Avocado Production in Michoacán, Submission
ID: SEM-23–002, CEC, July 24 2023, www.cec.org/submissions/registry
-of-submissions/avocado-production-in-michoacan.

10. Interviews in this chapter with Cuauhtémoc Sáenz-Romero, David
James, Andy Davis, and Karen Oberhauser are from Monika Maeckle,
"Should We Give Up Avocados to Help Save the Monarch Butterfly
Roosting Sites?" Texas Butterfly Ranch, Februrary 8, 2023, https://
texasbutterflyranch.com/2023/02/08/should-we-give-up-avocados-to
-help-save-the-monarch-butterfly-roosting-sites.

11. Columba González-Duarte, "Butterflies, Organized Crime, and 'Sad
Trees': A Critique of the Monarch Butterfly Biosphere Reserve Program
in a Context of Rural Violence." *World Development* 142 (June 2021),
105420, https://doi.org/10.1016/j.worlddev.2021.105420; phone interview
with Columba González-Duarte, February 22, 2023.

Chapter 15

1. "Climate Change and the Monarch Butterfly Migration Symposium,"
panel discussion, Texas Butterfly Ranch, October 20, 2016, http://
texasbutterflyranch.com/2016/11/21/climate-change-and-the-monarch
-butterfly-migration-symposium-tackles-tough-questions.

2. Cuauhtémoc Sáenz-Romero et al., "*Abies religiosa* Habitat Prediction
in Climatic Change Scenarios and Implications for Monarch Butterfly
Conservation in Mexico," *ScienceDirect*, July 2012, www.fs.fed.us/rm
/pubs_other/rmrs_2012_Sáenz_romero_c001.pdf.

3. Phone interview with Jerry Rehfeldt, February 21, 2023.

4. According to the UNAM Nahuatl dictionary, *oyametal* derives from *oya*,
to strip or shell some something, and *metl*, which can mean maguey
cactus or stone hand grinder such as a metate. www.gdn.unam.mx
/diccionario/consultar/palabra/oya/id/59029.

5. Arnulfo Blanco Garcia interview, March 2017.

6. "Southern North America: Southern Mexico," World Wildlife Fund, www.worldwildlife.org/ecoregions/nt0310; J. E. Fa and L. M. Morales, "Patrones de Diversidad de Mamíferos de México," in *Diversidad Biológica de México: Orígenes y Distribución*, ed. T. P. Ramamoorthy, R. Bye, A. Lot, and J. Fa, 315–54 (Mexico City, Instituto de Biología, UNAM, 1993).

7. Isabelle Aubin et al., "Why We Disagree about Assisted Migration: Ethical Implications of a Key Debate Regarding the Future of Canada's Forests," *Forestry Chronicle* 87, no. 6 (December 2011), https://pubs.cif-ifc.org/doi/10.5558/tfc2011-092.

8. "A Modern Ark," *Economist*, November 26, 2015, www.economist.com/special-report/2015/11/26/a-modern-ark.

9. Kara Rogers, *Quiet Extinction, Stories of North America's Rare and Threatened Plants* (Tucson: University of Arizona Press, 2015), 107–10.

10. Alejandea Borunda, Craig Welch, Sarah Gibbens, and Andrew Curry, "4 Solutions for Trees and Forests Threatened by a Hotter World," *National Geographic*, April 14, 2022, www.nationalgeographic.com/magazine/article/solutions-fixing-forests-fight-warming-feature?loggedin=true&rnd=1677015189753.

11. Stephen Buranyi, "How British Columbia Is Moving Its Trees and Why Other Provinces Have Yet to Follow Suit," *Motherboard*, January 20, 2016, https://motherboard.vice.com/en_us/article/how-british-columbia-is-moving-its-trees-assisted-migration-larch; Emma Marris, "Forestry: Planting the Forest of the Future," *Nature* 459 (June 17, 2009), 906–8, https://doi.org/10.1038/459906a.

12. Katharine Hayhoe,"Climate Change and the Monarch Butterfly Migration Symposium," panel discussion, Texas Butterfly Ranch, October 20, 2016, http://texasbutterflyranch.com/2016/11/21/climate-change-and-the-monarch-butterfly-migration-symposium-tackles-tough-questions.

13. Interview by phone with Sáenz Romero, February 16, 2023.

14. Suzanne Simard, *Finding the Mother Tree: Discovering the Wisdom of the Forest*, 1st ed. (New York: Knopf, 2021), 165.

15. "Colonia de Mariposas Monarca al Pie del Popocatépetl," Milenio video, February 4, 2018, www.youtube.com/watch?v=MYtmtWRH06w.

Chapter 16

1. "Administrator Pruitt Promotes Environmental Stewardship, Innovation, and Partnership at Commission for Environmental Cooperation," CEC Press release, June 27, /2018, www.epa.gov/archive/epa/newsreleases /administrator-pruitt-promotes-environmental-stewardship-innovation -and-partnership.html.

2. "Damage to Monarch Butterfly Colonies in 2016 Storm Worse Than Thought," *Science Daily*, September 17, 2017, www.sciencedaily.com /releases/2017/09/170917201441.htm.

3. Patrick Barkham, "Butterflies: Out of the Blue," *Guardian*, July 18, 2010, www.theguardian.com/environment/2010/jul/18/large-blue-butterflies -conservation.

4. Steven Morris, "Large Blue Butterfly Enjoys Best UK Summer on Record," *Guardian*, September 19, 2018, www.theguardian.com/environment /2018/sep/19/uk-large-blue-butterfly-best-summer-record.

5. Interview with Wayne Wehling, January 4, 2019.

6. Email and phone exchanges with Andy Davis, Chip Taylor, Carols Galindo Leal, Anurag Agrawal, and Karen Oberhauser, January 2019.

7. Walmart US Patent Application Publication 2018/0065749 A1, March 8 2018, Systems and Methods for Pollinating Crops via Unmanned Vehicles.

8. Statista, "Net Sales Share of Walmart U.S. in the United States in Fiscal Year 2022, by Merchandise Category," March 20, 2023, www.statista.com /statistics/252678/walmarts-net-sales-in-the-us-by-merchandise-unit.

9. Daniel Boffey, "Robotic Bees Could Pollinate Plants in Case of Insect Apocalypse," *Guardian*, October 9, 2018, www.theguardian.com /environment/2018/oct/09/robotic-bees-could-pollinate-plants-in-case -of-insect-apocalypse.

10. Scott Hoffman Black and Eric Lee-Mäder, "Can Robobees Solve the Pollination Crisis?" *Wings* 41, no. 1 (Spring 2018), 3, https://xerces.org /wings-magazine/spring-2018-grasslands.

11. "Are Robotic Bees the Future?" Goulson Lab, U.S. University of Sussex, www.sussex.ac.uk/lifesci/goulsonlab/blog/robotic-bees.

12. Interviews with Michael Martin, Clark McLeod, Anurag Agrawal, Sonia Altizer; see Monika Maeckle, "Monarch Zones: Bold Move or Bad Idea

for Expanding Monarch Butterfly Population?" Texas Butterfly Ranch, May 17, 2018, https://texasbutterflyranch.com/2018/05/17/monarch-zones -bold-move-or-bad-idea-for-expanding-monarch-butterfly-population.

Chapter 17

1. iNaturalist started in 2008 as a U.C. Berkeley School of Information Master's final project of Nate Agrin, Jessica Kline, and Ken-ichi Ueda.
2. Scott Weidensaul, "A Matter of Timing: Can Birds Keep Up with Earlier and Earlier Springs?" *Audubon Magazine*, Spring 2022, www.audubon .org/magazine/spring-2022/a-matter-timing-can-birds-keep-earlier-and.
3. Louie Bond, "Snow-Mageddon: Looking back at Winter Storm Uri, One Year Later," *Texas Parks and Wildlife Magazine*, January-February 2022, https://tpwmagazine.com/archive/2022/jan/ed_3_snowmageddon/index .phtml?fbclid=IwAR2wGJ2HqX1c6d8kMtcZbIXcSeFJuW7gwzud4Xi5w hfDrKhxhWisgWz7uDs.
4. Ellen Sharp, "Early Departure" *Journey North*, March 10, 2021, https:// journeynorth.org/monarchs/news/spring-2021/03102021-spring -migration-begins.
5. Chip Taylor, "Nectar Plants Used by Monarchs during March in Texas," Monarch Watch blog, May 25, 2021, https://monarchwatch.org/blog/2021 /05/25/nectar-plants-used-by-monarchs-during-march-in-texas.
6. Monika Maeckle, "Milkweed Shortage Sparks Alternative Fuels for Hungry Monarch Caterpillars," Texas Butterfly Ranch, April 11, 2014, https://texasbutterflyranch.com/2014/04/11/milkweed-shortage-sparks -alternative-fuels-for-hungry-monarch-caterpillars.
7. Lincoln Brower,"Ecological Chemistry," *Scientific American* 220, no. 2 (February 1969), 22–29, www.scientificamerican.com/article/ecological -chemistry.
8. Maeckle, "Milkweed Shortage Sparks Alternative Fuels."
9. Claire Fahy, "California's Monarch Butterflies Are Down 99%: Can This Plant Help?" *New York Times*, June 1, 2021, www.nytimes.com/2021/06 /01/science/butterfly-habitat-california.html.
10. David G. James, "Western North American Monarchs: Spiraling into Oblivion or Adapting to a Changing Environment?" Department of

Ecology, Washington State University, March 16, 2021, www.degruyter
.com/document/doi/10.1515/ami-2021–0002/html?lang=en.

11. Helena Horton, "Red Admiral Butterfly Population Soars 400% in
UK as Winters Warm," *Guardian*, August 3, 2023, www.theguardian
.com/environment/2023/aug/03/red-admiral-butterfly-population-uk
-sightings-winters-warm-climate?fbclid=IwAR3Saq9X_Xdnsbw-ajcpi
sdqS8P4VLr52ouczdZIpwJkswzJOa5M1B5ddtI.

12. Monika Maeckle, "Monarch Butterfly Population: Volatile, Unpre-
dictable, with a Ray of Hope," Texas Butterfly Ranch, January 27,
2022, https://texasbutterflyranch.com/2022/01/27/monarch-butterfly
-population-volatile-unpredictable-with-a-ray-of-hope.

13. David James and Teresa A. James, "Migration and Overwintering
in Australian Monarch Butterflies (*Danaus plexippus* (L.) (Lepidoptera: Nymphalidae): A Review with New Observations and Research
Needs," *Journal of the Lepidopterists' Society* 73, no. 3 (September 2019),
https://bioone.org/journals/The-Journal-of-the-Lepidopterists-Society
/volume-73/issue-3/lepi.73i3.a7/Migration-and-Overwintering-in
-Australian-Monarch-Butterflies-Danaus-plexippus-L/10.18473/lepi
.73i3.a7.short.

Chapter 18

1. Interview with Lincoln Brower, February 15, 2015; Monika Maeckle,
"Q & A: Dr. Lincoln Brower Talks Ethics, Endangered Species, Milkweed and Monarchs," Texas Butterfly Ranch, February 15, 2015, https://
texasbutterflyranch.com/2015/02/16/q-a-dr-lincoln-brower-talks-ethics
-endangered-species-milkweed-and-monarchs.

2. Natasha Daly, "Monarch Butterflies Are Now an Endangered Species," *National Geographic*, July 21, 2002, www.nationalgeographic
.com/animals/article/monarch-butterflies-are-now-an-endangered
-species.

3. Andy Davis, "I Am Officially Challenging the IUCN Listing of North
American Monarchs," MonarchScience blog, May 13, 2023, www
.monarchscience.org/single-post/i-am-officially-challenging-the-iucn
-listing-of-north-american-monarchs.

4. Chip Taylor, "Why There Will Always Be Monarchs: Reproductive Rate, Replacement, Resilience and Extinction," Monarch Watch blog, August 25, 2023.

5. Monika Maeckle, "Recent IUCN 'Endangered' Listing Creates Confusion for Monarch Butterfly Fans," Texas Butterfly Ranch, July 26, 2022, https://texasbutterflyranch.com/2022/07/26/recent-iucn-endangered -listing-creates-confusion-for-monarch-butterfly-fans.

6. IUCN Red List categories and criteria, version 3.1, 2nd. ed., https://www .iucnredlist.org/resources/categories-and-criteria.

7. IUCN summary sheet of endangered listing criteria, https://www .iucnredlist.org/resources/categories-and-criteria.

8. Email exchange with Chip Taylor, March 28, 2023.

9. Email exchange with Monica Bohm, March 27, 2023.

10. David G. James et al., "First Population Study on Winter Breeding Monarch Butterflies, *Danaus plexippus* (Lepidoptera: Nymphalidae) in the Urban South Bay of San Francisco, California," *Insects* 12, no. 10 (October 12, 2021), http://doi.org/10.3390insects 12100946.

11. In 2018, section 650 of California's Fish and Game code was updated to include terrestrial invertebrates. It applied to species on California's list of Terrestrial Invertebrates of Conservation Priority, which as of 2021 includes monarchs.

12. Gail M. Morris, Christopher Kline, Scott M. Morris, "Status of *Danaus plexippus* in Arizona," *Journal of the Lepidopterists' Society* 69, no. 2 (2015), 91–107, www.swmonarchs.org/Top%20Ten%20Findings%20of%20 Status%20of%20Danaus%20plexippus%20in%20Arizona.pdf.

13. Interview with Gail Morris, April 4, 2023.

14. Hannah B. Vander Zanden, "Alternate Migration Strategies of Eastern Monarch Butterflies Revealed by Stable Isotopes," *Animal Migration* 5 (2018): 74–83, www.degruyter.com/document/doi/10.1515/ami-2018–0006 /html.

15. Andy Davis interviewed by Monika Maeckle, "Monarch Butterfly Population up in Mexico, down in California." January 17, 2019, https:// texasbutterflyranch.com/2019/01/17/monarch-butterfly-population-up -in-mexico-down-in-california.

Afterword

1. Monika Maeckle, "Snout-Nosed Butterfly Population Surges in South Texas Following Recent Rains," Texas Butterfly Ranch, August 3, 2018, https://texasbutterflyranch.com/2018/08/03/rain-following-drought -brings-return-of-snout-nosed-butterflies-to-south-texas/#.

2. Mike Quinn, "South Texas Snout 'Migration' Ecology," Texasento.net, October 12, 2009, www.texasento.net/snout.htm#:~:text=South%20 Texas%20Snout%20%22Migration%22%20Ecology,reoccurring%20 south%20Texas%20entomological%20events.

▼ ▼ ▼

FURTHER READING

Agrawal, Anurag. *Monarchs and Milkweed: A Migrating Butterfly, a Poisonous Plant, and Their Remarkable Story of Coevolution.* Princeton University Press, 2017.

Halpern, Sue. *Four Wings and a Prayer: Caught in the Mystery of the Monarch Butterfly.* Vintage, 2002.

Hayhoe, Katharine, *Saving Us: A Climate Scientist's Case for Hope and Healing in a Divided World.* Atria/One Signal, 2021.

Mikula, Rick. *The Family Butterfly Book.* Storey, 2000.

Simard, Suzanne. *Finding the Mother Tree: Discovering the Wisdom of the Forest.* Knopf, 2021.

Tallamy, Doug W. *Nature's Best Hope: A New Approach to Conservation That Starts in Your Yard.* Timber Press, 2020.

Urquhart, Fred A. *The Monarch Butterfly: International Traveler.* Burnham, 1987.

Monarch Joint Venture website, https://monarchjointventure.org.

MonarchScience blog, www.monarchscience.org.

Monarch Watch website, https://monarchwatch.org.

Pollinator Partnership website, www.pollinator.org.

Xerces Society website, https://xerces.org.

INDEX

Italicized references indicate illustrations.